PAINTING AN
DECORATING

An Information Manual

AN LEA NY LIBRARY

Fourth Edition

A. Fulcher   B. Rhodes   W. Stewart
D. Tickle   J. Windsor

070188

© Padim Technical Authors
1975, 1981, 1989, 1998

Blackwell Science Ltd
Editorial Offices:
Osney Mead, Oxford OX2 0EL
25 John Street, London WC1N 2BL
23 Ainslie Place, Edinburgh EH3 6AJ
350 Main Street, Malden
    MA 02148 5018, USA
54 University Street, Carlton
    Victoria 3053, Australia
10, rue Casimir Delavigne
    75006 Paris, France

Other Editorial Offices:

Blackwell Wissenschafts-Verlag GmbH
Kurfürstendamm 57
10707 Berlin, Germany

Blackwell Science KK
MG Kodenmacho Building
7-10 Kodenmacho Nihombashi
Chuo-ku, Tokyo 104, Japan

First published in Great Britain by
Granada Publishing Ltd in Crosby Lockwood
Staples 1975
Reprinted 1976, 1979
Reprinted by Granada Publishing in 1980
Second edition 1981
Reprinted 1982
Reprinted with minor revisions 1983
Third edition published by BSP Professional Books, 1989
Reprinted 1990 (twice), 1991, 1993
Reprinted by Blackwell Science 1994, 1995, 1996
Fourth edition published 1998
Reprinted 1999

Set in 9.5/11pt Times
by DP Photosetting, Aylesbury, Bucks
Printed and bound in the United Kingdom
at the University Press, Cambridge

DISTRIBUTORS

Marston Book Services Ltd
PO Box 269
Abingdon
Oxon OX14 4YN
(*Orders*: Tel: 01235 465500
          Fax: 01235 465555)

USA
Blackwell Science, Inc.
Commerce Place
350 Main Street
Malden, MA 02148 5018
(*Orders*: Tel: 800 759 6102
                781 388 8250
          Fax: 781 388 8255)

Canada
Login Brothers Book Company
324 Saulteaux Crescent
Winnipeg, Manitoba R3J 3T2
(*Orders*: Tel: 204 837 2987
          Fax: 204 837 3116)

Australia
Blackwell Science Pty Ltd
54 University Street
Carlton, Victoria 3053
(*Orders*: Tel: 03 9347 0300
          Fax: 03 9347 5001)

A catalogue record for this title
is available from the British Library

ISBN 0-632-04159-5

Library of Congress
Cataloguing-in-Publication Data

Painting and decorating: an information manual/
A. Fulcher ... [et al.]. – 4th ed.
    p. cm.
    Includes index.
    ISBN 0-632-04159-5 (pbk.)
    1. House painting.   2. Painting, Industrial.
    3. Interior decoration.   I. Fulcher, A. (Alf)
IN PROCESS   98–55554
698'.1–dc21   CIP

For further information on
Blackwell Science, visit our website:
www.blackwell-science.com

CONTENTS

Preface **vi**

PART ONE   **Tools and Equipment**   **1**

1.1 Small tools and equipment **3**
1.2 Stripping and cleaning **32**
1.3 Spraying **38**
1.4 Power sources **57**

Table 1.1 The construction of a painter's brush **13**
Table 1.2 Signwriting brush sizes **26**
Table 1.3 Grades of blast cleaning **36**
Table 1.4 Sizes of spray gun orifice **39**
Table 1.5 Pressure drop related to hose size **47**
Table 1.6 Fluid-tips in common use with decorative paints **53**
Table 1.7 Air delivery/air pressure requirement of equipment **57**
Table 1.8 Properties and uses of butane and propane gases **61**

PART TWO   **Surface Coatings and Specialist Materials**   **65**

2.1 Introduction **67**
2.2 Primers, undercoats and stoppers **79**
2.3 Finishing paints **86**
2.4 Special paints **92**
2.5 Clear finishes and wood treatments **101**
2.6 Paint defects **107**
2.7 Gilding and bronzing materials and defects **112**
2.8 Stencil and masking materials **116**

Table 2.1 Properties of pigments **70**
Table 2.2 Properties of extenders **71**
Table 2.3 Properties of drying oils **72**
Table 2.4 Properties of natural resins **73**
Table 2.5 Properties of synthetic resins **73–74**
Table 2.6 Properties of oil-modified resins **75**
Table 2.7 Properties of other film-forming substances **76**
Table 2.8 Defects of surface coatings **107–10**
Table 2.9 Defects of paint in the tin **111**
Table 2.10 Types of gold leaf **113**
Table 2.11 Other metallic leaves **113**
Table 2.12 Gilding defects **115**

PART THREE   **Preparation of Surfaces for Paint and for Clear Finishes**   **127**

3.1 Abrasives **129**
3.2 Making good **132**
3.3 Preparation of previously painted surfaces **135**
3.4 Preparation of new, stripped or untreated surfaces **138**
3.5 Natural wood finishes **149**
3.6 Surface and paint film defects **151**

Table 3.1 Properties of coated abrasives **129–130**
Table 3.2 Comparison of grades of abrasive **130**
Table 3.3 Properties of abrasive powders and other abrasives **131**
Table 3.4 Properties of stoppers **133**
Table 3.5 Properties of fillers **134**
Table 3.6 Comparison of effects of liquid paint-remover and heat-stripping **136**

3.7   Preparation of surfaces for graining,
        marbling and surface texturing   **156**
3.8   Preparation of surfaces for gilding and
        bronzing   **157**

Table 3.7   Types of washing material   **137**
Table 3.8   Properties of building materials   **139–140**
Table 3.9   Non-ferrous metals – surface preparation and
                primers   **147**
Table 3.10   Galvanic series   **147**
Table 3.11   Surface defects   **151–152**
Table 3.12   Paint film defects   **153–155**

PART FOUR   Surface Coverings   159

4.1   Preparation of surfaces for surface
        coverings   **161**
4.2   Adhesives   **163**
4.3   Preparatory papers   **165**
4.4   Flat papers   **167**
4.5   Embossed papers   **169**
4.6   Relief papers   **171**
4.7   Washable papers   **172**
4.8   Decorative laminates   **174**
4.9   Defects of surface coverings   **178**
4.10 Glossary of terms   **182**
4.11 International performance
        symbols   **185**

Table 4.1   Preparation of bare surfaces   **161**
Table 4.2   Preparation of previously treated
                surfaces   **162**
Table 4.3   Defects due to surface or room conditions   **178**
Table 4.4   Defects due to paste and pasting   **179**
Table 4.5   Defects due to hanging   **180**
Table 4.6   Defects arising in manufacture   **180**
Table 4.7   Summary of common defects and their
                causes   **181**
Table 4.8   International performance symbols   **185**

PART FIVE   Scaffolding   187

5.1   Ladders, trestles, stepladders   **189**
5.2   Boards, staging, working
        platforms   **194**
5.3   Scaffolding   **196**
5.4   Suspended scaffolds   **205**
5.5   Ropes, knots   **208**
5.6   Safety equipment   **210**
5.7   Statutory obligations   **213**

PART SIX   Safety, Health and Welfare   215

6.1   Health and safety at work   **217**
6.2   Fire precautions   **221**
6.3   Storage of materials   **226**
6.4   Lead paint   **229**
6.5   Hygiene   **230**
6.6   Fumes, dust, noise, vibration   **232**
6.7   Lifting   **237**
6.8   First aid   **238**

Table 6.1   Fire extinguishers   **224**
Table 6.2   Storage of materials   **226–228**
Table 6.3   Causes of industrial dermatitis   **231**
Table 6.4   Protective equipment   **234–235**

PART SEVEN Specification 243

7.1   Components of buildings –
      exterior 245
7.2   Components of buildings –
      interior 246
7.3   Components of buildings –
      staircases 247
7.4   Types of door and window 248
7.5   Job planning 250
7.6   Choice of method of paint
      application 253
7.7   Costing 255
7.8   Calculations 258
7.9   British Standards 261

Table 7.1   Comparison of methods of paint
            application 253–254
Table 7.2   Principal units used in painting and
            decorating 259
Table 7.3   Calculating the surface areas of common
            shapes 260

PART EIGHT Colour 263

8.1   Colour and light 265
8.2   Colour organisation 267
8.3   Colour standards 270
8.4   Terms related to colour schemes 274

Table 8.1   Basic identification colours for pipelines 270
Table 8.2   Safety signs 272
Table 8.3   Safety colours 273

PART NINE Glazing 277

9.1   Types of glass and glazing
      compound 279
9.2   Handling and storage of glass 281
9.3   Measuring and cutting glass 282
9.4   Methods of glazing 283
9.5   Double glazing 284

Table 9.1   Grades of clear sheet glass 279

INDEX 285

PREFACE

This book is designed on the principle of a series of concise and simply written technical information sheets, providing essential job knowledge and understanding related to practical training schemes for the painter and decorator. It does not give instruction in practical skills, but has been written by a group of practising lecturers of painting and decorating in a way that they consider useful to augment training and further education courses in the craft.

The content has been planned to cover the Knowledge Evidence requirements of NVQs and the related knowledge elements of the City and Guilds 594 syllabus. As well as for NVQ and City and Guilds students, the book should also be of value to BTEC students studying construction, and to anyone requiring an up-to-date knowledge of painting materials and processes.

The Fourth Edition of *Painting and Decorating* has been updated by removing materials and practices which have become obsolete and introducing new materials, tools and plant which seem likely to become a permanent feature of the industry. It also now contains details of the common construction documents which were missing from previous editions and are needed to meet the requirements of current NVQ knowledge evidence. The major change is the addition of rapidly developing health, safety and welfare legislation, in particular the CDM Regulations, and the removal of references to obsolete regulations where appropriate. The section on British Standards has also had a major revision to bring it up to date, including appropriate European standards.

Consideration was also given to aligning the text more closely to NVQ knowledge evidence, but given the constant development and frequent changes to NVQs, it was decided that this would be unhelpful. Nevertheless, the book should prove of considerable use to NVQ students throughout their training.

The authors wish to thank the many specialists who have given generous assistance with this and the previous three editions, and to thank the following for advice and permission to reproduce their own material: ICI Paints for plates 1–8 and 11–14; Timber Research and Development Association for plates 9 and 10; Wallcovering Manufacturing Association for the wallcovering symbols; and the BSI for material from various standards.

PART ONE
Tools and Equipment

1.1 Small tools and equipment — 3

1.1.1 Painters' hand tools — 3
1.1.2 Paint containers — 8
1.1.3 Mechanical hand tools — 8
1.1.4 Burning-off equipment — 11
1.1.5 Brushes: components and types — 13
1.1.6 Paint rollers — 17
1.1.7 Paint pads and mittens — 19
1.1.8 Paperhangers' tools and equipment — 19
1.1.9 Graining, marbling and broken colour tools and brushes — 22
1.1.10 Signwriting and decorative painting tools and brushes — 26
1.1.11 Gilding tools and brushes — 29
1.1.12 Texture painting tools — 30

1.2 Stripping and cleaning — 32

1.2.1 The steam stripper — 32

1.2.2 Flame-cleaning — 33
1.2.3 Blast cleaning — 34

1.3 Spraying — 38

1.3.1 The spray gun — 38
1.3.2 Containers for spraying — 42
1.3.3 Compressors, receivers, transformers — 44
1.3.4 Hoses and connections — 46
1.3.5 Types of spray system — 48
1.3.6 Airless spray — 51
1.3.7 Cleaning, maintenance and storage of spray equipment — 54
1.3.8 Spray guns for special purposes — 55

1.4 Power sources — 57

1.4.1 Compressed air — 57
1.4.2 Electricity — 58
1.4.3 Gases — 60
1.4.4 The internal combustion engine — 61

1.1 SMALL TOOLS AND EQUIPMENT

1.1.1 PAINTERS' HAND TOOLS

Fig. 1.1   *Construction of a good-quality painter's knife*

Stripping knife or scraper (Fig. 1.2)
Blades must be capable of taking and retaining a good edge.
Size   25 mm, 38 mm, 50 mm, 68 mm, 75 mm, 100 mm wide.
Use   To remove old wallpaper, old paint films or loosely attached deposits.

Fig. 1.2   *Stripping knife*

Filling knife (Fig. 1.3)
Similar to a stripping knife in appearance but having a thinner gauge, specially treated metal blade which is more flexible. The blade need not be sharp, but must be thin and true, without any nicks.
Size   Up to 150 mm wide.
Use   To apply filler to open-grain timbers and small holes or shallow indentations in uneven surfaces.
Care   When not in use protect blade edge with a timber or soft aluminium cover.

Fig. 1.3   *Filling knife*

Caulking tool (Fig. 1.4)
Rigid plastic blade set in a wooden or plastic handle.
Size   200 mm, 250 mm, and 300 mm wide.
Use   For general filling and bedding-in scrim tape along plasterboard or fibreboard joints before texturing, or other forms of direct decorating onto dry lined surfaces.

Fig. 1.4   *Caulking tool*

Palette knife (Fig. 1.5)

Long narrow flexible blade with rounded end. Should be so balanced that when laid on the bench the blade will not touch the bench top.

Size Blades from 75 to 300 mm long.

Use For mixing paint both in tins and on palette boards.

Care Ensure that the end does not become bent or burred over.

Fig. 1.5 *Palette knives*

Stopping knife, putty or glazing knife (Fig. 1.6)

Usually has one side of the blade straight and the other curved, but can be obtained with a double curved blade.

Size Blade length of either 112 mm or 125 mm.

Use
 (i) To force putty and hard stopper into small holes and cracks.
(ii) To bevel facing putties when glazing.

Care If the point becomes worn or burred over, the blade should be reground.

Fig. 1.6 *Stopping knife*

Shave hook (Fig. 1.7)

The head has a bevelled cutting edge round the perimeter and is available in the three shapes illustrated.

Use
 (i) To scrape paint debris from mouldings, cornices and ornamental beadings, usually in conjunction with paint removers or burning-off equipment.
(ii) To cut out cracks in plaster surfaces prior to stopping.

Care Maintain sharp edge with a file.

Fig. 1.7 *Shave hooks*

Universal scraper (Fig. 1.8)

Short sharp replaceable blade attached to a large handle.

Size Blades between 45 and 80 mm wide.

Use To remove old paint, varnish or discolouration from timber surfaces.

Care Edge must always be kept sharp.

Fig. 1.8 *Universal scraper*

Hacking knife (Fig. 1.9)

A heavy, rigid metal blade with a leather handle. The back of the blade is flat to take hammer blows.

Size Blades either 100 mm or 125 mm long.

Use To remove old hard putties before reglazing.

Fig. 1.9 *Hacking knife*

Paint stirrer (Fig. 1.10)
A stiff blade with holes cut along its length. The flat end fits the bottom of the tin and paint flows through the holes, mixing the paint more quickly.
Size Various sizes available up to 600 mm long.
Use To stir paint.

Fig. 1.10 *Paint stirrer*

Hammer (Fig. 1.11)
Size 6 oz (170 g) to 8 oz (227 g).
Use
 (i) With nail punch, chisel and hacking knife.
 (ii) To drive in nails, sprigs and pins.
(iii) To remove heavy rust scale.

Fig. 1.11 *Hammers*

Pincers (Fig. 1.12)
Size 150 mm, 175 mm or 200 mm.
Use To remove nails and glazing sprigs.

Fig. 1.12 *Pincers*

Nail punch (Fig. 1.13)
Size 2 mm, 3 mm, 5 mm point sizes.
Use To punch nail-heads below the surface in woodwork before stopping.

Fig. 1.13 *Nail punch*

Measuring tape (Fig. 1.14)
Coated linen, PVC-coated fibreglass or steel tape.
Size From 2 to 30 m long.
Use To measure for estimating.

Fig. 1.14 *Measuring tape*

Straight-edge (Fig. 1.15)
Straight wooden batten with bevelled edge and two blocks to lift from surface.
Size 300 mm to 1 m long.
Use To guide a lining fitch when running lines (see also Fig. 1.92).

Fig. 1.15 *Straight-edge*

Screwdriver (Fig. 1.16)
Size Blade length about 125 to 150 mm.
Use To remove fitments, window and door furniture before painting and paperhanging.

Fig. 1.16 *Screwdriver*

Wire brush (Fig. 1.17)

 (i) Hardened and tempered steel wires in a wooden handle.

(ii) Bronze wires, which will not cause a spark, for use in areas with a high fire risk.

Size Available in a variety of shapes and sizes from 65 to 285 mm long.

Use

 (i) To remove corrosion from iron and steel.

(ii) To remove loose deposits from surfaces before painting.

Fig. 1.17   *Wire brushes*

Pointing trowel (Fig. 1.18)

Size 125 mm or 150 mm blade.

Use To make good large cracks and holes.

Fig. 1.18   *Pointing trowel*

Wood chisel (Fig. 1.19)

Size Short blade approximately 20 mm wide.

Use

 (i) To remove old putties, in place of a hacking knife.

(ii) For general purposes during surface preparation.

Fig. 1.19   *Wood chisel*

Paint strainer (Fig. 1.20)

 (i) Metal: tinplate, galvanised iron or zinc. The replaceable copper-gauze straining discs are available in three grades, 30-mesh coarse, 40-mesh medium and 60-mesh fine. Must be cleaned thoroughly, immediately after use, to prevent the mesh becoming clogged.

 (ii) Cardboard frame and muslin mesh disposable strainer.

(iii) Nylon cloth or fine muslin stretched over the kettle.

Use To remove dirt and skins from paint and varnish.

Fig. 1.20   *Paint strainers*

Filling board and hawk (Fig. 1.21)

Made of oiled plywood, laminated plywood, metal-covered plywood or thick-gauge plastic.

Size For stopping, approximately 100×130 mm; for fillers, approximately 180×230 mm – plus the length of the handle.

Use To mix and hold stoppers and fillers before and during application. The hawk is used to hold plaster or sand and cement when filling in large holes and cracks.

Fig. 1.21   *Filling board and hawk*

Rubbing block (Fig. 1.22)
Made of wood, plastic or rubber (the last-named has the longest life).
Size A rubbing face of approximately 70 mm wide × 100 mm long.
Use To hold abrasive papers, ensuring a flat, smooth surface and providing a more comfortable grip.

Fig. 1.22   Rubbing block

Sponge filling tool (Fig. 1.23)
Circular shaped synthetic sponge set on a flat stock at right angles to a plastic handle.
Size 200 mm diameter.
Use For smoothing the filling compound used in the caulking of plasterboard joints.

Fig. 1.23   *Sponge filling tool*

Glasscutter (Fig. 1.24)
Single wheel, six wheel, and diamond cutters are available. The hardened steel wheel types are most commonly used and are lubricated during use by dipping in white spirit. The six wheel type is in a revolving head to enable blunted wheels to be changed easily.
Use For scribing or scoring the glass at the point where it is to be cut.

Fig. 1.24   *Glasscutter*

Straight-edge (Fig. 1.25)
In use the glasscutter is guided along a straight-edge of varying lengths and made of wood or plastic.
Use Long straight-edges require steadying hands at both ends and glasscutting becomes a two person operation. When positioning the straight-edge, an allowance of 3 mm is made for the width of the cutting head of the glasscutter.

Tee-square (Fig. 1.25)
A straight-edge of wood or plastic with a tee-piece guide at one end set at right angles to the straight-edge.
Use The guide fits the edge of the glass to ensure a cut at right angles to the edge.

Fig. 1.25   *Straight-edge and T-square*

Glazing piers (Fig. 1.26)
Metal pliers with flattened ends to grip the glass.
Use To grip the glass when cutting thin strips. The pliers apply gentle leverage to break the glass along the scored cut made by the glass-cutter.

Fig. 1.26   *Glazing pliers*

1.1.2 PAINT CONTAINERS

Paint kettle or can (Fig. 1.27)
Made of galvanised sheet iron, black sheet steel,
zinc, aluminium alloy or plastic.
Size Available with 125 mm, 150 mm, 180 mm,
200 mm diameter, or holding approximately $\frac{3}{4}$, 1,
$1\frac{1}{2}$, $2\frac{1}{2}$ litres of paint.
Use To hold a convenient quantity of paint
decanted from manufacturer's container.
Care Clean out thoroughly with appropriate
solvent immediately after each job. Aluminium
kettles will dissolve if cleaned with caustic soda.
Burning-out weakens welds and joints. Heat and
strong solvents may destroy plastic kettles.

Fig. 1.27 *Paint kettle and kettle hook*

Kettle hook or pot hook (Fig. 1.27)
Wire hook with a double prong, one end used to
suspend the kettle from the ladder rung, leaving
the operative with both hands free.

Bucket or pail (Fig. 1.28)
Made of galvanised iron, rubber or plastic.
Size 7 litres, 9 litres, 14 litres.
Use To hold water, washing solution, paste or
water-thinned paints.
Care Plastic buckets will melt if subjected to a
naked flame.

Fig. 1.28 *Buckets*

Paint tray (Fig. 1.29)
 (i) Rectangular metal or plastic tray; various
     sizes to take rollers from 150 to 350 mm
     wide.
(ii) Tank, bucket or scuttle with either a raised
     side or a wire grid. Several sizes available
     containing up to 10 litres of paint.
Use To hold paint for roller application,
designed to ensure an even take-up of paint.

Fig. 1.29 *Paint trays*

1.1.3 MECHANICAL HAND TOOLS

Rotary disc sander (Fig. 1.30)
A disc sander has a flexible rubber or composition sanding-head, attached to a motor which
can be driven either by electricity or by compressed air. On to the sanding-head can be fitted
any one of a range of abrasive papers.

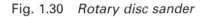

Fig. 1.30   *Rotary disc sander*

Use
 (i) Fitted with abrasive disc, for general pre-
     paration of surfaces including joinery, floor
     surfacing, previously painted surfaces and
     for removal of rust. Suitable for work on
     curved surfaces.
(ii) A lambswool mop can be fitted over the
     sanding head and used for polishing.
(iii) A carborundum grinding wheel can be
     fitted to remove rough surfaces from welds.
Note   Difficult to control and liable to damage
the surface or leave it uneven.
Safety precautions   See 'Rotary wire brush'.

Rotary wire brush (Fig. 1.31)
Radial cup-or disc-shaped wire brush which can
be fitted to air-driven and electrically operated
motors.
Use   To remove loose and flaking paint and
rust from metal surfaces.
Safety precautions
 (i) Goggles must always be worn.
(ii) Hand-held power tools with exposed
     rotating heads must be switched off and
     stopped before they are laid down; they
     may otherwise spin, causing damage and
     injury.
(iii) Portable abrasive wheels larger than 55 mm
     diameter must be marked with maximum
     speed specified by the manufacturer.
(iv) In explosive atmospheres phosphor-bronze
     brushes must always be used.

Fig. 1.31   *Rotary wire brush*

Orbital sander (Fig. 1.32)
Electric or air-driven sander consisting of a
rectangular flexible platform pad on to which
various types of abrasive paper are fixed. The
platform moves in a small circular or orbital
motion which abrades the surface. The rate of
orbit varies according to make and type but can
range from approximately 6000 to 20 000 orbits
per minute: usually, the faster the motion the
better the finish. Orbital sanders are comparatively light in weight and can be used for long
periods without undue fatigue.
Use   To prepare and smooth timber, metal,
plastic and previously painted surfaces.

Fig. 1.32   *Orbital sander*

Safety Electrically operated types can constitute a safety hazard if used with water. Air-driven types are perfectly safe.

Note Although slower in use they are easier to control and produce finer surfaces than rotary sanders.

Belt sander (Fig. 1.33)

A sanding machine with a continuous belt of abrasive paper. It maintains a flat sanding action but abrades at a faster rate than orbital sanders.

Size Belt sizes either 75 or 100 mm wide × 610 mm long. Other large heavy-duty machines are available for sanding floors.

Use

 (i) To sand large areas of timber.

(ii) To sand metal surfaces to remove light rust.

Fig. 1.33   *Belt sander*

Needle gun or descaling pistol (Fig. 1.34)

An outer body or casing containing a number of hardened steel needles which are propelled forward by an air-driven spring-loaded piston. On hitting the surface the needles rebound and are again forced forward by the piston. This action is continuous, working at the rate of approximately 2400 blows per minute. The individual needles are self-adjusting, thus ideal for use on uneven surfaces.

Use For the removal of rust, particularly around awkward areas such as nuts and bolts. The method is too slow for economic use on large areas. Also used for cleaning stonework and ornamental ironwork.

Power Driven by compressed air with an air consumption of approximately 5 to 10 cfm (2.36 to 4.72 litre/s) according to the size of the gun, at 90 psi (6.1 bar).

Needles Three types are available:

 (i) Pointed tip for breaking through heavy rust and millscale. These leave a rough surface.

(ii) Chisel-ended, similar use to pointed tip but less damaging to the surface, leaving only a slight surface-mark.

(iii) Flat-ended. These will not mark the metal surface and can be used on thin-gauge metals. Also used on surfaces which require a light treatment, such as concrete aggregate and stonework.

Fig. 1.34   *Needle gun*

Safety precautions Goggles must be worn when descaling. *Never* use in an explosive atmosphere – sparks cause explosions (special spark-resisting needles can be obtained for this work).

Descaling hammer and chisel (Fig. 1.35)
This can be a conversion of the needle gun or a specially designed pneumatic or electrically-operated descaling hammer. It works by a piston moving backwards and forwards at high speed inside a cylinder. The piston produces hammer blows on the chisel.
Use The constant hammering action of the chisel is used to remove heavy rust and scale. The process is relatively slow and not economic on large areas. The surface of the metal may be damaged as the chisel leaves pits and burrs which are difficult to coat with a full paint system.
Safety precautions Goggles must always be worn when descaling.

Fig. 1.35   *Descaling chisels*

Fig. 1.36   *Bottle-type gas torch*

1.1.4 BURNING-OFF EQUIPMENT

Heat is used to soften paint films sufficiently to allow the paint to be scraped off with stripping knives and shave hooks. Must not be used to remove lead paint systems – see 6.4.2.

LPG gas torch
(a) *Bottle-type torch* (Fig. 1.36)
Operated by liquefied petroleum gases (LPG), either butane or propane (see 1.4.3). Portable and light-weight, with refill gas bottle which screws under the torch. Various nozzles can be fitted with different jets to produce a range of flame shapes and varying degrees of heat. A small bottle of gas can last from 2 to 4 hours depending on the type of nozzle used.

If this type of lamp is operated at an acute angle it may flare up or stop.
(b) *Independent torch* (Fig. 1.37)
Burners attached to hoses which are fixed to large gas cylinders containing from 10 lb (4.5 kg) to 32 lb (14.5 kg) of propane or butane gas. Two torches can be run from one cylinder. They are lighter to use and provide easier movement than the bottle type, particularly in awkward spaces.

A *interchangeable burner*
B *trigger locking ring*
C *trigger*
D *regulator*
E *refill cylinder (propane or butane)*

Fig. 1.37   *Independent gas torch*

(c) *Disposable cartridge type* (Fig. 1.38)
Burner or torch head screwed into a disposable
gas cartridge. Very light to use but more
expensive than the returnable 'refill' bottles.

These torches have a short burning-time and
the flame produces less heat than the larger
types.

Mains gas torch or burner (Fig. 1.39)
A hand-torch operated by connection to natural
or coke gas supply, convenient in the workshop
if gas supply is laid on but with limited use on
sites.

Hot air paint stripper (Figs. 1.40 and 1.41)
Similar in many ways to an electric hair drier.
Hot air is produced by an electric filament which
can be easily adjusted to any temperature
between 20°C and 600°C.

With no flame the risk of fire is much less than
with blow-lamps and gas torches. There is less
chance of cracking glass and it is very difficult to
scorch timber.

Ideal for use on old or delicate surfaces, and in
areas of high fire risk such as eaves of old
buildings.

Safety precautions
(i) *Always* use recommended reinforced hose
     to avoid kinking and damage on site and to
     withstand effect of LPG.
(ii) *Always* ensure that curtains and soft fur-
     nishings are removed from the area before
     burning off.
(iii) *Always* make sure that no woodwork is left
      smouldering before leaving the job.
(iv) *Always* keep a bucket of sand or a fire
     extinguisher at hand to smother any fire
     immediately.
(v) *Always* make sure all charred and burnt
     paint is extinguished before placing it in a
     metal bin (safer to place all debris in a
     bucket of water).
(vi) *Do not use* to remove old lead paints (see
     6.4.2).

A *burner*
B *expendable gas
  container*

Fig. 1.38   *Disposable cartridge gas torch*

Fig. 1.39   *Mains gas torch*

heater

blower

Figs 1.40 and 1.41   *Hot air paint strippers*

A *handle*
B *stock*
C *setting*
D *filling*

Fig. 1.42 *The construction of a painter's brush*

1.1.5 BRUSHES: COMPONENTS AND TYPES

TABLE 1.1 **The construction of a painter's brush** (Fig. 1.42)

A	**Handle**	Usually of hardwood such as beech, ash or elm, sealed to make handling and cleaning easier and to stop water soaking into the wood. May also be made of plastic.
B	**Stock**	The means by which the handles are fixed to the fillings: usually a ferrule made from nickel-plated polished steel, but a few are made of copper. Two types: (a) sheet metal riveted to the handle; (b) seamless type pressed on to the handle.
C	**Setting**	An adhesive which cements the filaments together at the root. Main types used are vulcanised rubber or a synthetic resin, usually epoxy.
D	**Filling**	Four main types: (a) bristle; (b) man-made fibres such as nylon, Perlon, Tynex or Orel; (c) natural fibres; (d) mixtures of bristle, hair and fibres, may also be referred to as *union*. 'Length out' is the amount of filling visible.

Characteristics of fillings
Pure bristle (Fig. 1.43)
This term applies only to hairs from the pig, hog or boar. As domesticated pigs have only a small amount of short, soft hair, the bristles used in paint brushes come from wild pigs, hogs or boars found mainly in China and India. Chinese bristles are of high quality, very resilient and black in colour. Indian bristles are rather coarse, longer in length, less resilient than Chinese bristles and they vary in colour. The main properties of bristle are:
(i) Strength and resilience: they can have a long length out yet still work well and hold the weight of paint.

(ii) A split tip or *flag* which provides a fine soft tip for laying off.

(iii) Serrations or scales like minute teeth along the length of the bristle: these prevent the hairs lying too close together, allowing the brush to hold more paint.

(iv) A natural taper from root to tip which allows a bundle of bristles to lie together in a brush form.

(v) A natural curl or lean: this enables the bristles to be curved inwards and lie together. Dusting-brushes are reversed and curve outward.

Horsehair (Fig. 1.44)

Obtained from the manes and tails of horses. This hair has little spring and no taper, but can be mechanically split or flagged. It is sometimes mixed with bristle to produce a softer and cheaper brush. When mixed with pure bristle this must be clearly stated on the brush.

Orel, nylon, Tynex, Perlon (Fig. 1.45)

These are all synthetic 'man-made' filaments produced to imitate bristle. They can be tapered and flagged but they do not have serrations, and cannot therefore hold paint as well as bristles. They are very springy and recovery is equal if not superior to bristle. They are more hard-wearing, not affected by ordinary solvents, resistant to many chemicals and not damaged by insects or fungi. Have become more widely used because of the scarcity and increasing cost of bristles.

Fibre (Fig. 1.46)

Obtained from grass and plants, mainly from Mexico. They have no taper and no serrations, but can be flagged mechanically. They are coarse, with very poor resilience. Used as an adulterant with bristles or horsehair to reduce cost or used alone in washing-down brushes. Not attacked by alkali.

Furs and soft hairs

Very fine hairs used in specialist brushes for signwriting, graining and gilding.

Sable   This term applies to hairs cut from the tail of various animals of the weasel family. The most commonly used is from the Kolinsky. The best hairs are obtained from animals which live in the wild Arctic regions of Asia, principally China.

Sable is a red-brown hair. Although fine, the hair has great strength and will spring back into shape once pressure upon it has been released. It is tapered and produces a very fine tip that

Fig. 1.43   *Pure bristle*

Fig. 1.44   *Horsehair*

Fig. 1.45   *Synthetic filament*

Fig. 1.46   *Fibre*

makes it ideal for fine, controlled painting. Used principally for signwriting brushes.

Ox   Hair cut from the ear of an ox. They are a little darker in colour than sable and although equally fine, they do not have the strength or springiness of sable. Long lengths used in coach liners. When mixed with sable they may be called 'Sabox' fillings. Used as a substitute for sable to save costs.

Squirrel   Still incorrectly called camel hair and obtained from the blue squirrel, generally those native to the colder regions of Asia and America. It is a dark brown hair and much softer than sable. Squirrel is used as the filling in gilders tips, sword stripers, mops, and graining cutters.

Badger   The most commonly available hairs are obtained from the Chinese badgers. It is long hair, generally off-white in colour with a brown stripe in the centre. It has the peculiar characteristic of being slightly thicker in the middle than at its fine tip. This property produces fillings which swell out towards the end, quite opposite to most brush filling which tend to taper. Badger hair is used in softeners.

Synthetic filaments Most commonly made from nylon. They are very fine hairs with considerable strength and springiness. They can be produced in any colour although bright red-brown is a common colour used. Principal manufacturer is Japan. Used in signwriting brushes and one-strokes.

Flat paint or varnish brush (Fig. 1.47)

Usually pure bristle but also available in man-made filaments. Quality varies according to the amount and quality of the filling and the 'length out'.

Size 25 mm, 37 mm, 50 mm, 62 mm, 75 mm, 100 mm wide. Also varying in thickness and length of filling.

Use To apply most types of paint and varnish to a wide variety of surfaces including doors, sashes, frames and wall areas.

Fig. 1.47 *Flat paint brush*

Flat wall brush (Fig. 1.48)

Available in wide range of varying quality including pure bristle, man-made filament and a mixture of bristle, hair and fibre (union). Quality is also dependent on weight of filling used and the length out.

Size 100 mm, 125 mm, 150 mm, 175 mm wide.

Use

(i) To apply emulsion paints to large areas.

(ii) To apply adhesive to surface coverings.

Fig. 1.48 *Flat well brush*

Two-knot brush (Fig. 1.49)

Available mainly in pure bristle or union. The knots are usually bound in copper.

Size Varies by weight of filling used and length out.

Use For same purpose as flat wall brush.

Fig. 1.49 *Two-knot brush*

Washing-down brush (Fig. 1.50)

Cheaper type of two-knot brush usually union or fibre, bound in copper wire.

Size Usually available in one size only.

Use

(i) To wash down and to apply caustic solution.

(ii) To apply water-thinned paints to rough surfaces such as brick and stucco.

(iii) To apply cement-based paints (as the filling is not attacked by alkalis).

Fig. 1.50 *Washing-down brush*

Dusting brush (Fig. 1.51)

Black or white filling of either pure bristle or man-made filaments, usually nylon.

Fig. 1.51 *Dusting brush*

Size   Available as standard type of brush or as a three-or four-ring type, both 90 or 110 mm wide.
Use   To remove dust and dirt from surfaces before painting.

Fitch (Fig. 1.52)
Bristle filling, usually white, set in either a round or a flat nickel ferrule.
Size   Flat type from 5 to 28 mm wide. Round type from 3 to 20 mm in diameter. Available in eight sizes.
Use   To apply paint to detailed work and areas difficult to reach with paint brush.

Fig. 1.52   *Fitches*

Radiator or flag brush (Fig. 1.53)
Bristle filling attached to a long wire handle which can be bent to fit into awkward areas.
Size   25 mm, 37 mm, 50 mm wide.
Use   To apply paint to restricted spaces such as behind pipes and column radiators.

Fig. 1.53   *Radiator brush*

Crevice brush or bent fitch (Fig. 1.54)
Similar to fitch, but angled to assist working in awkward spaces.
Size   Flat or round, in sizes from 9 mm, 12 mm, 15 mm wide and 18 mm, 31 mm in diameter.
Use   Similar to radiator brush.

Fig. 1.54   *Crevice brush*

Cement paint brush (or block brush)
(Fig. 1.55)
Man-made filling or coarse white fibre, set in polished wooden handle.
Size   150 mm wide.
Use   An inexpensive brush for the application of cement paint, and other masonry finishes. The tough filling resists the abrasive action of application to rough surfaces like cement rendering and brickwork. The filling is also unaffected by the lime content of cement paints.

Fig. 1.55   *Cement or block brush*

Care and maintenance of brushes
Before using a new brush, flirt the bristles to remove any loose hairs.
Overnight storage   Brushes used in oil or alkyd paints can be stored overnight by suspending in water or placing in a patent 'brush keep'.
Cleaning brushes after use   Clean thoroughly to remove all traces of paint, especially from the root where paint tends to accumulate and harden, causing the brush to fan out. *Method:* Remove all traces of paint with a suitable solvent. Wash out the solvent in warm detergent solution, then rinse thoroughly in clean water. Dry the brush by hanging in a well-ventilated room. Brushes stored damp will be ruined by mildew.
Storage Wrap the fillings in waterproof paper and lay flat in boxes. Sprinkle with moth-repellent. Store should be neither too warm nor too dry (about 18°C).
Never stand a brush on the bristles as this can damage it; either lay brushes flat or suspend them. *Never* wash or use a pure bristle brush in alkaline material as they destroy the bristle.

1.1.6 PAINT ROLLERS

Types of roller
Cylinder roller (Fig. 1.56 a and b)
This is the standard type of roller used by the
decorator. It consists of a straight cylinder or
core with a fabric cover, also called a sleeve. The
cylinder is fitted to an axle which is attached to a
handle. Most types have replaceable coverings
to save on the cost of new axles and handles.

The handles are fitted either (a) with a single
arm frame which makes it easier to apply paint
close to corners; or (b) with a double arm which
ensures a more even pressure over the width of
the roller. This is especially useful when used
with an extension handle, but cannot be used
close to corners.

(a)   (b)

Fig. 1.56   *Cylinder rollers*

Curved roller (Fig. 1.57)
Produced with a hollow section, cut to allow a
wrap-round on pipes and poles.

Another type of roller designed for painting
pipes and poles has a number of separate roller
sections fitted to a spring axle. The axle bends to
almost any diameter and the roller is loaded in
an ordinary tray.

Rollers are also produced with several hollow
sections designed for painting corrugated
sheeting.

Fig. 1.57   *Curved rollers*

Airless-fed roller (Fig. 1.58)
A cylinder roller fed by an airless spray tip. Paint
is pumped through the handle from an airless
spray pump. The roller is instantly charged and
the paint flow controlled by a trigger on the
handle.

Application is fast but this type of roller has a
limited use because it requires an airless spray
pump to feed paint to the roller.
Advantages
  (i) Can be used on contracts where application
      by spray is not allowed.
 (ii) No overspray.
(iii) Reduces need for masking.
(iv) Extension arms can be fitted to provide a
      reach of nearly 2 m. This enables reason-
      ably high ceilings and walls to be coated
      without using a scaffold.

Fig. 1.58   *Airless-fed roller*

Extension arms
These can be used with most rollers to enable
reasonably high ceilings and wall areas to be
coated without using a scaffold.

Types of covering
Natural materials
Mainly pelts from the sheep (lambswool) and the goat (mohair). Wool varies in density, thickness and general properties with every animal, also according to species and country of origin. Therefore roller coverings made from these materials can vary a great deal.

Lambswool, long pile   Holds a good quantity of paint, ideal for use on heavy textured surfaces such as brick, concrete and pebble-dash. The long pile tends to cling together and become matted and it leaves a heavy texture on smooth surfaces.

Lambswool, short pile   Up to 10 mm is available. This makes a good general-purpose roller, ideal for water-thinned and oil paints on smooth surfaces.

Mohair   Has a short pile, spreads paint easily, leaves very little texture and the pile does not become matted. Ideal for gloss paints and for use on flat surfaces.

Man-made filaments – Dynel, Nevel, Nylon, Dacron These are all hard-wearing man-made filaments, some of which are designed specifically for paint-roller coverings. They are manufactured within controlled limits and can be produced with all the properties desirable in a paint-roller covering, i.e.:
  (i) to hold a reasonable quantity of paint;
  (ii) to release the paint to the surface easily and apply a full coat;
  (iii) to leave the paint film smooth and even, with minimum texture;
  (iv) to be able to get into surface irregularities;
  (v) to have a length of filament (pile) suitable for a certain type of paint and type of job. Roller fabrics are available with pile lengths ranging from 6 to 32 mm.

Short pile (up to 10 mm long)   Good general-purpose roller, used to apply emulsions and synthetic oil paints to smooth and textured surfaces such as plaster, plasterboard, hardboard and smooth metal.

Medium pile (up to 20 mm long)   General-purpose roller for smooth, medium and rough surfaces such as embossed papers, fairfaced concrete and brickwork. Ideal for water-thinned paints.

Long pile (up to 32 mm long)   Specially made for very rough surfaces such as the application of oil-and water-based masonry paints to pebble-dash and rough-cast.

Synthetic plastic foam
Very cheap and often thrown away after each job. When fully charged the rollers tend to skid and apply the paint unevenly.

Size
Standard cylinders are generally available from 150 to 450 mm wide.

Curved rollers are available in various sizes up to 200 mm wide. The spring-axle type is adjustable to various diameters.

Use
Rollers can be chosen from the wide range available to suit most types of work including large areas of flat surface, anaglypta and ingrain papers, rough-cast, brickwork and concrete. They are also very useful on perforated hardboard (pegboard) acoustic boards, wire netting, corrugated sheeting and pipework.

Advantages
  (i) Paint is applied several times faster than with a brush on flat and textured surfaces.
  (ii) Chances of 'flashing' and 'sheariness' are reduced.
  (iii) Extension poles can reduce the need for scaffolding.
  (iv) No brush-marks, although a slight 'orange peel' texture is unavoidable.

Disadvantages
  (i) Rollers cannot paint right into angles and corners.
  (ii) They are not suitable for fine, intricate work.
  (iii) Some people object to the texture.

Care of rollers
  (i) Rollers must be clean and free from solvents before storing.
  (ii) After cleaning and washing, rollers should always be hung up to dry: never left resting on the pile, as this will crush and distort the covering.
  (iii) If stored damp, mildew can ruin lambswool or mohair coverings and the metal parts will corrode. Always store in clean, dry, well-ventilated conditions.

1.1.7 PAINT PADS AND MITTENS

Paint pad (Fig. 1.59)
A short pile face, usually mohair, attached to a foam pad or cushion which is fitted to a handle. Modified versions may have
(a) detachable head to allow for replacements, thus reducing costs;
(b) wheels inserted in the side of the plastic body to make cutting-in easier.

Size Various sizes are available depending on the manufacturer. They range from large (150 × 100 mm) down to a small sash-painter sometimes called a 'toothbrush' pad. Some firms also make a crevice pad.

Use Can be used to apply gloss, eggshell, flat and emulsion paints, wood stains and varnishes. Used on flat surfaces such as ceilings, walls, woodwork, building boards and over firmly attached wallpaper. Useful for painting bands or lines.

Care Clean only in water or white spirit and soap and water, as appropriate. Strong solvents and some types of brush cleaner may damage the foam pad. When clean, hang up to dry.

Fig. 1.59 *Paint pads*

Paint mitten (Fig. 1.60)
A mitten, made of sheepskin, designed to fit the hand like a normal glove-mitten.

Use Used to apply emulsion paint, primer, undercoats and finishes to surfaces such as iron railings, behind radiators and pipes which are difficult to coat by normal methods.

Care Clean in similar way to rollers.

Fig. 1.60 *Paint mitten*

1.1.8 PAPERHANGERS' TOOLS AND EQUIPMENT

Paperhanging brush (Fig. 1.61)
Made of white or grey bristle or synthetic filaments set in rubber.

Size 200 mm, 225 mm and 250 mm wide.

Use To apply papers and fabrics to walls and ceilings.

Care Keep clean: wash in soap and tepid water and hang to dry.

Fig. 1.61 *Paperhanging brush*

Scissors or shears (Fig. 1.62)
Made of polished steel with hollow-ground blades.

Size 250 mm, 275 mm and 300 mm long.

Fig. 1.62 *Scissors or shears*

Use To cut lengths and trim at angles and around obstacles.

Care Keep clean and sharp. Never clean with abrasive paper as this rounds off the cutting edge. Store dry and clean to prevent rusting.

Fig. 1.63   *Seam rollers*

Seam and angle roller (Fig. 1.63)

Made of hardwood. Side arm or double arm fitting.

Size 25 mm and 37 mm wide.

Use To roll down edges of paper at seams and angles.

Care Keep clean and lubricate sparingly.

Felt roller (Fig. 1.64)

Made of a number of felt discs or a felt-covered cylinder.

Size 90 mm and 175 mm wide.

Use In place of paperhanger's brush, it applies a firm, even pressure to the paper without distorting or polishing it.

Care Wash regularly in warm water.

Fig. 1.64   *Felt roller*

Casing wheel (Fig. 1.65)

Plain or serrated blades.

Size 37 mm wheel.

Use An alternative method to scissors and trimming knives for trimming surplus paper at angles or around obstacles.

Care Keep clean and sharp.

Fig. 1.65   *Casing wheels*

Hand trimmer (Fig. 1.66)

Two small cutting wheels attached to a guide which is held against the edge of the paper.

Use To remove the selvedge from untrimmed paper, or to cut narrow strips of paper. It is adjustable and can cut strips up to approximately 75 mm from the edge of the paper.

Care Replace blades when blunt.

Fig. 1.66   *Hand trimmer*

Paste board (Fig. 1.67)

Made from wood, usually collapsible for easy transportation. Should be firm.

Size 1.830 m long × 560 mm wide.

Use For measuring, cutting, matching and pasting surface coverings.

Care Keep face and edges clean and free from paste.

Fig. 1.67   *Paste board*

Rule (Fig. 1.68)
One metre long, made of boxwood and folded into four sections. Working is more accurate if the rule is used on its edge.
Use For measuring areas, lengths and widths.

Fig. 1.68 *Rule*

Trimming knife (Fig. 1.69)
Available with retractable or replacement blades, which are either angled or rounded.
Use For trimming and cutting at angles and around obstacles.
Care Keep edge sharp or replace blade regularly.

Fig. 1.69 *Trimming knife*

Steel straight-edge (Fig. 1.70)
Perfectly straight with a bevelled edge.
Size 600 mm, 915 mm, 1.400 m and 1.830 m long.
Use To guide a knife when trimming papers and fabrics. Used over a zinc strip to protect the paste board and ensure a clean cut.

Fig. 1.70 *Steel straight-edge*

Ridgely track-trimmer (Fig. 1.71)
A combined 1.830 m metal purpose-made straight-edge and cutting wheel used over a zinc strip 75 mm wide × 1.830 m long.
Use To trim or cut widths of paper or fabrics before or after pasting.

Fig. 1.71 *Ridgely track-trimmer*

Spirit level (Fig. 1.72)
Available in metal or wood to check horizontal, or horizontal and vertical lines.
Size Many sizes, but 1 m long is the most useful when paperhanging.

Fig. 1.72 *Spirit level*

Plumb bob and line (Fig. 1.73)
A small weight suspended on a length of fine cord.
Use To check that the first length of surface covering on every wall is upright.

Fig. 1.73 *Plumb bob*

Rubber roller (Fig. 1.74)
Solid rubber roller.
Size   90 mm and 175 mm wide.
Use   To hang materials such as vinyl papers and photo murals where a felt roller or brush would not be heavy enough.
Care   Keep clean.

Fig. 1.74   *Rubber roller*

Spatula (Fig. 1.75)
Straight-edged flexible metal or plastic blade set in a plastic stock and handle.
Size   250 mm and 450 mm wide
Use
(i) For smoothing vinyl papers and fabric-backed vinyl wall coverings.
(ii) For pressing vinyls and papers into angles and acting as a guide when cutting with a trimming knife.

Fig. 1.75   *Metal spatula*

1.1.9 GRAINING, MARBLING AND BROKEN COLOUR TOOLS AND BRUSHES

Mottler (Fig. 1.76)
Pure white bristle filling, set in a metal ferrule. Sometimes has a short wooden handle.
Size   50 mm and 75 mm wide.
Use
(i) For simulating the soft lights and darks seen beneath the surface grain of many woods (mottling).
(ii) As a general graining brush to simulate the straight grain and heartwood patterns of mahogany, walnut, and similar woods.

Fig. 1.76   Mottler

Overgrainer (Fig. 1.77)
Pure white bristle filling, set in a metal ferrule. Similar to a mottler but with a longer filling, and much thinner in section.
Size   50 mm and 75 mm wide.
Use   To add fine parallel grain patterns to completed work (overgraining). Often used with water graining colour and passed through a coarse comb to separate the filling.

Fig. 1.77   *Overgrainer*

Pencil overgrainer (Fig. 1.78)
Separate tufts of pure white bristle filling, set in a metal ferrule. Also available with a pure sable hair filling but expensive to buy.
Size 50 mm and 75 mm wide.
Use Similar to those for the overgrainer.

Fig. 1.78 *Pencil overgrainer*

Fan fitch overgrainer (Fig. 1.79)
Separate tufts of pure white bristle filling, set in a flattened metal ferrule to produce a fan shape.
Size 25 mm and 38 mm at the widest end.
Use
  (i) To simulate the curly grain patterns of burr walnut, pollard oak, and similar decorative woods.
 (ii) Used as an overgrainer.

Fig. 1.79 *Fan fitch overgrainer*

Flogger (Fig. 1.80)
Long pure bristle filling, set in a wooden stock or a metal ferrule.
Size From 75 mm to 100 mm wide.
Use
  (i) To simulate the pore markings of wood by beating wet graining colour.
 (ii) To imitate the decorative texture 'dragging'.

Fig. 1.80 *Flogger*

Badger softener (Fig. 1.81)
Separate tufts of pure badger hair filling, set in a wooden stock. Some smaller varieties are set in a metal ferrule.
Size 75 mm and 100 mm wide. Smaller flat varieties: 25mm, 38mm, 50mm and 75mm wide. Smaller round varieties: small size No. 2, to large No. 12.
Use To gently fade-out or soften the sharp edges of patterns produced in graining and marbling. Although intended only for use with water based materials, they are often used with oil graining colour and marbling gilp.

Fig. 1.81 *Badger softener*

Hog hair softener (Fig. 1.82)
Similar to the badger softener, but containing a pure white bristle, (hog hair) filling.
Size 75 mm and 100 mm wide.
Use As for the badger softener but not as soft and intended for oil based materials only.

Fig. 1.82 *Hog hair softener*

Drag brush or brush grainer (Fig. 1.83)
A filling of pure bristle on one side and coarse nylon on the other, set in separate tufts in a wooden stock.
Size   100 mm wide.
Use   To simulate a variety of straight grain patterns by dragging through wet graining colour.

Fig. 1.83   *Drag brush or brush grainer*

Cutter (Fig. 1.84)
Short squirrel hair filling, set in a metal ferrule.
Size   50 mm and 75 mm wide.
Use   To simulate the fine clean highlights common in feather mahogany and similar decorative woods. Used in conjunction with a wet chamois leather to remove water graining colour in the area of the highlights.

Fig. 1.84   *Cutter*

Filbert fitch (Fig. 1.85)
Short pure white bristle or synthetic filling, set in a flattened metal ferrule.
Size   A range from small No. 0 to large No. 16
Use
 (i) For simulating heartwood grain patterns of oak, pine, and similar woods.
(ii) Painting marble veins.

Fig. 1.85   *Filbert fitch*

Steel combs (Fig. 1.86)
Flexible steel combs with long, parallel, square cut teeth.
Size   25 mm, 50 mm and 75 mm wide, with fine, medium, and coarse graded teeth.
Use
 (i) To simulate the coarse straight grain patterns of oak, pine, and similar woods. Used covered in clean linen rag, and dragged through wet graining colour. Finer grades used for 'cross combing' to break-up the coarse straight grain.
(ii) To produce the decorative texture 'combing' by dragging through wet scumble applied to walls or panels.

Fig. 1. 86   *Steel combs*

Rubber combs (Fig. 1.87)

Flexible rubber combs with short, vee-shaped teeth. Also available with square cut teeth.

Size Approximately 75 mm wide. Available with fine, medium, and coarse graded teeth. Triangular varieties contain all three grades of teeth arranged on the three edges. Also available with teeth graduated from fine to coarse on one edge.

Use

(i) To simulate the coarse straight grain patterns of oak, pine, and similar woods.

(ii) To imitate combing similar to steel combs.

Fig. 1.87   *Rubber combs*

Heart grain simulator (Fig. 1.88)

A moulded rubber appliance with a heartwood pattern engraved on the curved working face, and a flat toothed comb on the edges.

Size 75 mm and 125 mm wide, with fine, medium, and coarse grain patterns.

Use For the fast and repetitive simulation of heartwood patterns of oak, pine, and similar woods. The curved face is dragged with a rocking motion through wet graining colour. The comb is used to simulate the straight grain patterns at each side.

Fig. 1.88   *Heart grain simulator*

Check roller (Fig. 1.89)

A roller made from a series of loose fitting serrated metal discs. Sometimes equipped with a special clip-on mottler to feed graining colour to the discs, and to separate them in use.

Size 50 mm and 75 mm wide.

Use As a printing roller to simulate the dark broken pore markings of oak and similar woods.

Fig. 1.89   *Check roller*

Veining horn or thumb piece (Fig. 1.90)

A flexible strip of plastic, or any similar material shaped with one square end, and the other end rounded.

Size Approximately 100 mm long.

Use

(i) To simulate the grain patterns of quartered oak and heartwood of oak. Used with clean linen rag to 'wipe out' the patterns from wet graining colour.

(ii) For simulating the veins of some marbles.

Feathers

Swan, goose or other large fairly rigid feathers.

Use

(i) Applying marbling colour when producing either background textures or veins.

(ii) Producing wavy or curly grain markings of decorative timbers such as walnut.

1.1.10 SIGNWRITING AND DECORATIVE PAINTING TOOLS AND BRUSHES

Signwriting brushes (Fig. 1.91)

Also called writers, pencil brushes, or sables. Long handled brushes with sable, 'Sabox' or nylon fillings. They can be obtained:

Metal ferrule The most common form. The filling is enclosed within and attached to the handle by a cylindrical, seamless, non-rusting metal casing.

Quill Filling enclosed in the base of a bird's feather which is shrunk upon the handle.

Chisel end The tips of the filling finish to a flat end. Suitable for all types of signwriting.

Pointed end The tips of the filling finish to a point. Suitable for fine painting and lettering.

Sizes Metal ferrule brushes are available in sizes dependent on diameter and length-out. Quills are available as the bird's name which

Fig. 1.90 *Veining horn or thumb piece*

relates to the feather size from which they are made (see Table 1.2).

Use

(i) Signwriting and decorative painting.

(ii) Veining when marbling.

(iii) Heartwood painting when graining.

Fig. 1.91 *Signwriting brushes*

TABLE 1.2 **Signwriting brush sizes**

Ferrule	1	2	3	4	6	8	12
Quill	Crow	Small duck	Duck	Large duck	Goose	Ex Small swan	Condor
Diameter in mm	1.6	1.9	2.2	2.6	3.6	4.9	7.8

Lining fitch (Fig. 1.92)
Bristles set in a flat metal ferrule with a chisel edge cut at an angle.
Size   6 mm, 12 mm, 18 mm, 25 mm, 31 mm and 37 mm wide.
Use   Against a straight-edge, to paint straight lines.

Fig. 1.92   *Lining fitch*

Coach liners (Fig. 1.93)
Long haired brushes, usually with ox hair fillings. Available in quill of ferrule.
Sizes   Similar sizes to signwriting brushes.
Use   Running thin painted lines.

Fig. 1.93   *Coach liner*

Sword stripers (Fig. 1.94)
Long, dagger shaped squirrel hair fillings, either in a metal ferrule or bound to the handle with copper wire.
Sizes   In widths from 8 mm to 13 mm.
Uses   Running thin painted lines.

Fig. 1.94   *Sword striper*

Lining wheel (Fig. 1.95)
The cylindrical body is metal. Paint is poured in at one end when the plunger assembly is removed. The serrated wheel assembly controls the line thickness and is available in a number of wheel widths.
Size   Wheel sizes vary from 2 mm to 6 mm wide, and available in double wheel heads.
Use   To run thin painted lines, either against a straight-edge, or controlled by the guide which can be adjusted to slide against the edge of a board or panel. Can also be used to run curved lines.

Fig. 1.95   *Lining wheel*

One-stroke brushes (Fig. 1.96)
Metal ferruled brushes which have been squeezed flat to produce a spade-like filling. Commonly available with ox or 'sabox' fillings, but very expensive brushes are available with sable fillings.
Sizes   In widths from 4 mm to 25 mm.
Use
  (i) Painting types of script letters in one stroke, similar to that obtained with a broad nibbed pen.
 (ii) Filling-in large letters or large areas of decorative painting.
(iii) Painting-in heartwood grain when imitating oak, pine and similar woods.

Fig. 1.96   *One stroke brush*

Stencil brushes (Fig. 1.97)
The filling is of short stiff bristles.
Sizes Various sizes from 10 mm to 38 mm diameter.
Use To obtain decorative effects or lettering by applying paint through a cut-out stencil, or around a template.

Fig. 1.97 *Stencil brush*

Mahl stick (or rest stick) (Fig. 1.98)
A stiff rod of timber or metal padded at one end. Can be in one length, or in two or three lengths which screw or push together.
Size Approximately 500 mm long.
Use To steady the hand when signwriting or decorative painting.

Fig. 1.98 *Mahl stick*

Palette board or thumb board (Fig. 1.99)
A rectangular timber or plastic board which fits over the thumb of the non-brush hand.
Size Approximately 230 mm × 160 mm.
Use To hold dippers of paint and upon which to shape the brush when signwriting or decorative painting.

Fig. 1.99 *Palette or thumb board*

Dippers (Fig. 1.100)
Small cylindrical metal containers which clip onto the palette board.
Size Various, but most commonly used are 40 mm diameter × 40 mm high.
Use To hold paint and/or thinners when signwriting or decorative painting.

Fig. 1.100 *Dipper*

Stippling brush (Fig. 1.101)
Bristles set in small tufts in a flat stock. The tips of all the bristles are level. The handle may be straight or arched.
Size Various sizes from 100 mm × 75 mm to 200 mm × 150 mm.
Use
(i) To remove brush marks from wet scumble to produce a 'stippled' finish.
(ii) To 'blend' wet scumble.

Fig. 1.101 *Stippling brush*

Two-row stippler (Fig. 1.102)
A narrow version of a stippling brush consisting of only two rows of bristle tufts.
Size 100 mm × 25 mm and 150 mm × 25 mm.
Use In association with a stippling brush for stippling wet scumble in angles, and on narrow features such as mouldings and edges.

Fig. 1.102 *Two-row stippler*

Care and maintenance of signwriting brushes

Cleaning brushes after use Must be carried out immediately after use, otherwise paint which is allowed to harden at the root of the brush may never be removed completely without seriously damaging the hairs. Hardened paint will destroy the shape of the brush. The brush must be rinsed out in the thinners of the paint being used, e.g. white spirit for oil paints, and water for emulsions. When all traces of paint have been removed, the hair should be dried with clean rag to remove the thinners.

The final process is to shape the brush with grease. This is done by taking any non-setting grease or oil between the thumb and forefinger, and pushing it deep into the root of the filling, and pulling it up to the tip leaving the brush in its natural shape. The greasing protects the filling from damage while being carried around, as well as ensuring that any slight traces of paint which may have been left in the filling will not harden. The grease must be thoroughly washed out with white spirit before re-use.

The use of hot water or strong soaps and detergents to wash soft hairs may result in the hairs losing their springiness.

Storage Lay flat in a tray, drawer or brush box, or bind them to a stiff card using elastic bands. New brushes should be stored with an insecticide to protect them from moth attack which will either destroy the filling completely, or eat away the fine tip rendering the brush useless.

1.1.11 GILDING TOOLS AND BRUSHES

Gilders cushion (Fig. 1.103)
A flat board padded with felt and covered with chamois leather or goat skin. One end has a folded parchment shield to protect the gold leaf from draughts. Underneath the board are two loops, one for the thumb to hold the cushion secure, and the other to sheath the gilders knife.

Size Approximately 200 mm × 125 mm.
Use Upon which to cut loose gold leaf and other metallic leaves into convenient sizes when gilding relief surfaces.

Fig. 1.103 *Gilders cushion*

Gilders knife (Fig. 1.104)
Long flat bladed knife which is well ground but not sharp.

Size Usually a standard size of 125 mm blade length.
Use To cut loose gold leaf and other metallic leaves on a gilders cushion.

Fig. 1.104 *Gilders knife*

Gilders tip (Fig. 1.105)
Very thin row of squirrel or badger hair set between two pieces of card.

Size Usually 85 mm wide. Available in three lengths of hair: short up to 40 mm; medium up to 50 mm; and long about 60 mm length out.
Use When the filling is rubbed vigorously against the gilder's hair or cheek, the hairs become sufficiently charged with static elec-

Fig. 1.105 *Gilders tip*

tricity to attract the gold leaf from the cushion so that it may be applied to the sized areas.

Gilders mop (Fig. 1.106)

Domed shape squirrel hair with filling set in either a metal ferrule or quill.

Size A range of diameters from 6 mm to 25 mm.

Use
 (i) To push applied gold leaf into crevices when gilding relief surfaces.
 (ii) To 'skew' off surplus gold from relief surfaces.

Fig. 1.106   *Gilders mop*

1.1.12 TEXTURE PAINTING TOOLS

Fig. 1.107   *Mixing tool or bumper*

Mixing tool or bumper (Fig. 1.107)

A round plate made from rigid plastic or aluminium with holes near the rim. Takes a standard broom handle.

Use To ensure speedy and thorough mixing of texture paints by plunging up and down during the mixing operation.

Rubber stipplers (Fig. 1.108)

Type (a) Fine flexible rubber filling set into a flat base with an arched wooden or plastic handle.

Size 150 mm × 100 mm and 200 mm × 150 mm. A very large type measuring 330 mm × 330 mm is available which can be fitted with an extension handle for ceiling work.

Fig. 108   *Rubber stippler Type A*

Type (b) A more rigid stippler completely moulded in rubber with thicker filaments.

Size 150 mm × 60 mm.

Use To produce a range of stippled or swirl effects in texture paint.

Fig. 1.108   *Rubber stippler Type B*

Combs (Fig. 1.109)
Clear plastic flexible blade with 'V' grooves or curves along the edge. A shorter blade is set alongside to ensure even pressure when in use.
Size   90 mm, 180 mm, 250 mm and 300 mm wide.
Use   To produce a range of combed patterns in texture paint.

Fig. 1.109   *Comb*

Lacer (Fig. 1.110)
A triangular clear plastic blade with rounded corners, and a smaller blade attached to one corner.
Use   Used wet and drawn lightly over partially set texture paint to remove high points and sharp edges or spikes.

Fig. 1.110   *Lacer*

Textured rollers (Fig. 1.111)
Synthetic foam rollers with sculptured patterns.
Size   Similar to paint rollers approximately 200 mm long.
Use
 (i) Used over wet texture paint to give a variety of relief textures.
(ii) A smooth plastic type gives a timber bark effect and can also be used as a lacer.

Fig. 1.111   *Textured roller*

1.2 STRIPPING AND CLEANING

Fig. 1.112 *Electric steam stripper*

1.2.1 THE STEAM STRIPPER (Fig. 1.112)

There are two types:
(a) heated by liquefied gas; large, heavy piece of equipment.
(b) electrically heated; small, compact, easily moveable unit.

Steam strippers consist of a water-tank attached to an enclosed burner or heating element. The water is heated and converted into steam which, by means of a flexible hose, is carried to a perforated plate. When the plate is laid against the wall, steam penetrates deeply into the wall-covering, softening the paper and adhesive. Paper can then be easily scraped from the surface, leaving it smooth, clean and free from bacteria.

Use
(i) To assist in the removal of any type of surface covering from ceiling and walls. Particularly useful for materials which are difficult to remove by normal soaking, e.g. Lincrusta, washable, multi-layered papers and coated papers.
(ii) To assist in the removal of water paints, emulsions and some plastic paints.

Note
(i) If the concentrator is left in one place too long, steam may cause damage to the substrate, e.g. blistering of the top layer of plaster.
(ii) Not recommended for use on old or friable plaster surfaces as the steam penetration will damage the plaster.

Safety precautions
(i) Never let the water level drop too low.
(ii) Do not use kinked or tightly coiled hose which will block the flow of steam.

1.2.2 FLAME-CLEANING
(Fig. 1.113)

Flame-cleaning is a very effective method of removing rust from iron and steel. It will remove millscale also, providing it is not tightly adhering. A flame of extremely high temperature (up to 3000°C) is passed over the surface of the metal. Its effectiveness is due to several factors:
(i) *Differential expansion* When the intense localised heat is applied to partially weathered millscale, the scale, being heated first, expands faster than the base metal and flakes off.
(ii) *Dehydration* Rust is dehydrated by the intense heat. All the moisture is driven off rapidly and the rust is converted to dry powder which is easily removed by wire brush.
(iii) *Heat penetration* The heat of the flame penetrates the smallest cracks and pits in the metal and dries out all traces of moisture.

Equipment
Oxygen and acetylene gases are used together and burnt in about equal proportions. It is recommended that three cylinders of each gas are coupled together to reduce handling. By using a special propane mixer, propane gas can also be used.

Several different types of multi-jet nozzle are available: round, circular, semi-circular and flat. The flat heads are available from 50 to 200 mm wide. This range allows the equipment to be used around bolts, corners and in recesses as well as on flat areas.

Procedure
Normally three men work as a team: *Operative no. 1* flame-cleans the surface, which should appear light grey all over when finished; *Operative no. 2* wire-brushes the surface to move all the dry rust powder; *Operative no. 3* primes the metal while it is still warm (38°C, or the temperature at which the hand can be held comfortably on the steel). The warmth of the plate tends to lower the viscosity of the paint, allowing it to penetrate more easily into small surface irregularities so that an excellent bond is formed between primer and metal.

Use
The equipment is portable, therefore suitable for use on site – e.g. on bridges, railings, structural steel, gas-holders, ships' hulls – where rust and millscale have to be removed.

shank

100 mm flat   50 mm flat

round   circular

Fig. 1.113 *Flame-cleaning equipment*

Disadvantages
(i) The heat may buckle and distort steel plate less than 6 mm thick. Specialist advice must be obtained if working on thin plate.
(ii) When removing heavy millscale it is often more economical to chip the scale off before flame-cleaning.
(iii) It may not provide a sufficiently high standard of preparation for special coatings such as two pack epoxies.

Safety precautions
(i) All safety precautions necessary when using gases must be observed (see 1.4.3).
(ii) Leather gauntlets and tinted goggles should always be worn.
(iii) A face-mask is recommended for protection against fumes when old paint is being removed.

1.2.3 BLAST CLEANING

Blast cleaning, abrasive cleaning, shot-blasting, grit-blasting and abrasive blasting are all terms used to describe a very efficient operation whereby compressed air is used to shoot or blast various types of abrasive on to a surface at very high speeds (up to 450 mph or 725 km/h). The force of impact and the hardness of the abrasive pulverises the rust and roughens the surface of the material.

Use

(i)   To remove rust and millscale from iron and steel.
(ii) To clean surfaces, e.g. stone, concrete, alloy.
(iii) To obtain decorative effects on such materials as stone, concrete and glass.

Equipment

Two main types of equipment are available, with variations of each produced by different manufacturers.

(a) *Open blasting*

Used for approximately 90–95 per cent of all large-scale cleaning and surface preparation work.

Unit (Fig. 1.114)   The abrasive is loaded into a hopper and falls into a pressurised pot. It is then fed through a metering valve into the compressed-air stream through the hose to the nozzle.

The equipment is available in various sizes, the choice being mainly dependent on size and type of job.

Nozzle (Fig. 1.115)   The most important part of the equipment. Two types are available:

(i) Straight cylinder nozzle. Not very efficient and the cleaning pattern tends to be uneven.
(ii) Nozzle with a narrow straight section or throat inside the cylinder, known as a 'venturi tube'. This is much more efficient: velocity increases up to 450 mph (725 km/h) as the abrasive passes through the narrower section; the cleaning pattern is also very even.

The inner lining is of tungsten carbide or a special alloy to withstand the severe abrasive action.

Available in various lengths and orifice sizes.

Hose   The best hose is made of reinforced rubber containing carbon black for protection against static electricity.

Hose couplings (Fig. 1.116)   Hose couplings which fit internally are not recommended as they restrict the flow of abrasive. Special external couplings are made to allow a straight-through passage for the abrasive.

Air supply   The force of the abrasive blast is created by compressed air. The higher the air

air in

Fig. 1.114   *Open blasting equipment*

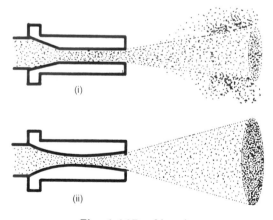

(i)

(ii)

Fig. 1.115   *Nozzle*

Fig. 1.116   *Hose couplings*

pressure, the greater the force of impact, which in turn produces a better finish. The ideal pressure is 100 psi (6.5 bar) *at the nozzle*. Less pressure than this means slower working and a poorer result.

Air consumption is high. It varies according to the type of equipment, nozzle and air pressure used, but is usually between 80 cfm (for small units) and 330 cfm (for larger plants) (37.8 and 156 litre/s). Quite large compressors are therefore required for this work (see 1.3.3 and 1.4.1).

(b) *Vacuum reclaim process* (Fig. 1.117)
The abrasive is fed from a storage tank to the blasting head, where it cleans the surface and is sucked back by vacuum into the plant. It is then screened and recirculated all in one operation. This is a relatively slow process, used mainly for small work and where cleanliness is of the utmost importance.

Special equipment is available which combines water or steam with the abrasive to reduce dust. This is effective, but slow and messy.

Abrasives
Choice of abrasive
It is very important to choose the correct type, size and shape of abrasive. The choice will depend on the job, and in each case the following points should be considered:
  (i) Type of surface to be cleaned, i.e. metal or stone.
 (ii) What has to be removed – heavy millscale, rust, old paint, etc.
(iii) Type of paint system to be used.
 (iv) Standards of finish required (see below, Standards).
  (v) Cost – whether the abrasive used is or can be reclaimed.

Types of abrasive
Re-usable abrasives Where it is possible and economical to reclaim and re-use abrasives (e.g. blast rooms, new steel storage tanks, vacuum units) mineral abrasives are used. These can often be re-used up to 10 times, depending on type. They include:
  cast steel – angular grit and round shot
  malleable and chilled iron – angular grit and round shot
  aluminium oxide
  aluminium
  brass } limited specialised use
  copper
  glass

Fig. 1.117 *Vacuum reclaim process*

Expendable abrasive On most site work it is difficult and expensive to recover the abrasives; cheaper, expendable types are therefore used once only. Mineral slags, produced in the refining of iron and copper, are mainly used. They should be hard and angular to give good cutting power.

Size of abrasive particles
Abrasives are available in a wide range of sizes, graded through a British Standard mesh and specified in BS 7079:1994. Selecting the correct size of abrasive for a particular job is very important. Large-particle grit or shot will hammer a surface harder and loosen scale faster than will small particles, but the resulting rough, coarse finish will not usually be suitable for painting. Small-particle abrasives, having more pellets to the pound, will produce a finer finish much more quickly, but will have less power of impact. After considering all these factors, for most work it is best to use the smallest size abrasive that will give the desired result for the work in hand.

Shapes of abrasive particle (Fig. 1.118)
Abrasives are obtainable in two shapes:
Shot Small round particles of abrasive which act as tiny hammers on the surface, producing a round profile.

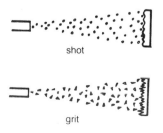

Fig. 1.118 *Abrasive particles*

TABLE 1.3 **Grades of blast cleaning**

BS 7079	SSPC (USA)	Swedish	Description
First quality	White metal	SA3	Perfectly white metal surface all over
Second quality	Near white metal	SA2.5	The whole surface shall show blast cleaning pattern and at least 95% of the surface shall be clean bare steel
Third quality	Commercial	SA2	The whole surface shall show blast cleaning pattern and at least 80% of the surface shall be clean bare steel

Note A fourth. 'Brush off' finish is sometimes used, but as this produces a much lower standard of preparation, it is not recommended or encouraged as suitable for painting.

Grit Small angular chips which act as tiny chisels on the surface, producing a sharp angular profile.

Standards

The standard to which a surface is prepared usually depends on all or some of the following factors:

(i) The type of paint system to be used.
(ii) The state of the metal before treatment.
(iii) The conditions to which the finished job will be subjected.
(iv) *Cost* – the higher the standard, the higher the cost.

Standards for blast cleaning are specified by the following bodies, of which the first two are internationally recognised:

The Swedish Standards Organisation (SIS 05 59 00)
The Steel Structures Painting Council, USA
The British Standards Institution (BS 7079:1994)

Note The Swedish standards are set by comparing the surface with a set of expertly prepared coloured photographs. The others are only verbal descriptions.

Three grades are commonly used and can be quoted with approximate equivalents in each standard (Table 1.3).

Profile (Fig. 1.119)

Some surface roughness is a great advantage in providing a physical key for the primer to adhere to; however, the degree of roughness should not be excessive, or peaks of steel may project through the paint film and rust very quickly. As a general guide, the roughness known as 'profile' or 'anchor pattern' should not exceed 50 microns.

Painting procedure

The surface should be painted as soon as possible after cleaning – never more than 4 hours later, or further corrosion will start before the

Fig. 1.119 *Profile*

surface is primed. (The time interval is dependent on weather and site conditions.)

Safety precautions
Abrasive cleaning can be dangerous if not carried out correctly. Used correctly, it is perfectly safe.
 (i) Always observe the correct working procedure which has been carefully drawn up by the manufacturer of the equipment used.

(ii) Always wear the recommended safety equipment:
Visor or blasting helmet to protect eyes and face.
Usually equipped with breathing apparatus.
Abrasive-resistant gloves.
Industrial safety boots.
Rubber or leather apron.

1.3 SPRAYING

1.3.1 THE SPRAY GUN (Fig. 1.120)

The spray gun set-up:
1 *Air cap or air nozzle*
  Directs air into the flow of paint, causing the paint to atomise and form a spray pattern
2 *Fluid-tip or fluid nozzle*
  Acts as a seating for the needle and controls the amount of paint flowing through
3 *Fluid needle*
  A spring-loaded needle controlled by the trigger which fits into the fluid-tip and starts and stops the flow of paint

Other components:
4 *Spreader adjustment valve*
  Regulates the amount of air flowing into the air cap, adjusting the fan pattern from a round to a wide fan
5 *Fluid adjustment valve*
  Controls the adjustment of the fluid needle, regulating the amount of paint flowing through
6 *Air valve*
7 *Trigger*
  Controls the air inlet valve and the fluid needle
8 *Material inlet*
  Where the paint enters the gun from either a suction pot or a fluid-hose from a pressure pot
9 *Air inlet*

air

paint

Fig. 1.120   *Components of the spray gun*

The set-up used on a spray gun depends on a number of factors, the first being the way in which the fluid and air leave the gun. Two types of gun are used for general spraying.

In this section bars and litres per second are used to express volumetric pressure because both units are commonly used by equipment manufacturers and the construction industry.

(a) Bleeder gun
A gun without an air valve: the air flows continuously through the gun and cannot be cut off. These guns are used with small compressors which have no air-receiver or pressure-controlling device. If the air were not allowed to bleed out, air pressure would build up in the hose and might damage the compressor. The trigger controls only the flow of paint.

TABLE 1.4   **Sizes of spray gun orifice**

Orifice size	Use	Air consumption (see section 1.4.1)
1.0 to 1.9 mm	Stains, etch primers, cellulose finishes, clear lacquers	3 to 7 cfm (1.4 to 3.3 litre/s)
1.4 to 2.2 mm	Most types of gloss, undercoat, emulsion and some primers	5 to 15 cfm (2.3 to 7 litre/s)
2.2 to 2.8 mm	Heavy-bodied paints, some texture paints and heavily pigmented fillers	7 to 16 cfm (3.3 to 7.5 litre/s)
2.8 to 9.5 mm	Heavy-texture paints, anti-condensation paints and bituminous compounds	17 to 25 cfm (8 to 11.8 litre/s)

Note   The amount of air consumed also depends on the air pressure used.

(b) Non-bleeder gun

A gun fitted with an air valve which shuts off the air when the trigger is released. The trigger therefore controls both the air and the flow of paint. Most spray guns are of this type.

The gun set-up

Although the gun set-up is usually considered to have three units, it can be more easily explained as having two:
(i) The fluid-tip and needle, which must go together and be a perfect fit.
(ii) The air cap or air nozzle – usually considered as a set with the fluid tip and needle, but some types can be interchanged.

(i) The fluid-tip and needle

These control the quantity of paint going into the air stream; the size used depends on:
(i) Type and viscosity of paint being sprayed.
(ii) Volume of air available.
(iii) Speed of application required.
(iv) Standard of finish required, fine or coarse atomisation.
(v) Type of air cap being used.

Table 1.4 provides an approximate guide to the range of orifice sizes for use with various types of paint.

(ii) Air nozzle or air cap

This is the most important part of the spray gun. It directs the air into the paint stream to break it up into a fine vapour (atomisation) and to form the various spray patterns. Choice of type depends on:
(i) Volume of air available.
(ii) Type and viscosity of paint used.
(iii) Type of material container used.
(iv) Size of fluid-tip and needle used.
(v) Size and type of object to be sprayed, e.g. large wall area, small intricate work such as ornamental ironwork.
(vi) Standard of finish required – fine or coarse atomisation.

Two basic types of air cap are available, the internal-mix and external-mix.

The internal-mix air cap (Fig. 1.121)   The air and fluid mix in a cavity inside the air cap before being released into the atmosphere. Used mainly with low-pressure equipment.

1 *air*
2 *needle*
3 *paint*

Fig. 1.121   *Internal-mix air cap*

Two types are available (Fig. 1.122):
(a) The paint sprays from a single slot in a fixed fan pattern.
(b) The paint sprays from a single round hole in a round pattern.

Use For work with compressors using $\frac{1}{4}$ or $\frac{1}{2}$ hp motors producing 2 to 3 cubic feet per minute or cfm (0.95 to 1.3 litre/s) – see 1.4.1.

Advantages of internal mix
(i) Requires low volume of air.
(ii) Requires low air pressure.
(iii) Has low overspray.

Disadvantages of internal mix
(i) Quick-drying paints tend to clog the exit holes.
(ii) Has coarse atomisation (see 1.4.1).
(iii) No control of the size or shape of the fan pattern.

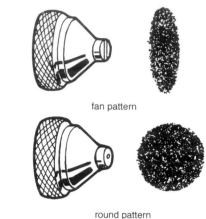

fan pattern

round pattern

Fig. 1.122   *Types of internal-mix air cap*

The external-mix air cap (Figs. 1.123 and 1.124) The most commonly used type: the air and paint mix outside the air cap.

There is a central orifice where the paint leaves the gun, surrounded by a circle of air. This effects the first stage of atomisation. Other holes around the central orifice (depending on the type) stop the pattern expanding too fast, help keep the horns clean and provide a further secondary atomisation.

The horns have holes or side port-holes which direct a stream of air from two sides, producing a fan pattern and providing additional atomisation.

1 *air*
2 *needle*
3 *paint*
4 *horns*

Fig. 1.123   *External-mix air cap*

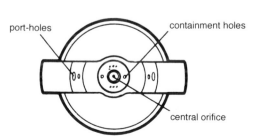

port-holes

containment holes

central orifice

Fig. 1.124   *Multiple-jet air cap*

Two types are available:

(a) *Syphon or suction-feed air cap* (Fig. 1.125) The fluid-tip extends beyond the air cap. A partial vacuum is produced as the paint leaves the fluid-tip (see 1.3.2). This type can also be used on gravity and pressure-feed guns.

1 *air*
2 *needle*
3 *paint*

Fig. 1.125 *Syphon-feed air cap*

(b) *Pressure-feed air cap* (Fig. 1.126) The fluid-tip and air cap are flush on this type, as no vacuum is required. The paint is forced to the gun under pressure.

There are between five and fifteen orifices in the air cap. The multiple jet air cap produces better atomisation and a more even fan pattern; good atomisation is possible with less air pressure, but does require greater volume of air.

Advantages of external mix
 (i) Very fine atomisation.
 (ii) Precise control over spray pattern size.
 (iii) Very wide range available.
Disadvantage of external mix
Higher volume of air required than with internal mix.

1 *air*
2 *needle*
3 *paint*

Fig. 1.126 *Pressure-feed air cap*

Air caps for special effects
Special heads or special set-ups to fit standard guns are available with orifice size and air cap modified to produce decorative effects such as veiling (Fig. 1.127), spattering (Fig. 1.128), flecking and shading or blending.

Although these are used with pressure-feed guns, most special effects require high fluid and low air pressures.

Note Spattering effects can be produced with a conventional set-up by increasing the pressure on the fluid and reducing the atomising air pressure. The result is an all-over spot pattern. The particle size may be adjusted by varying the pressures. This type of spraying demands a highly skilled operative to maintain an even, all-over pattern.

Fig. 1.127 *Veiling*

Fig. 1.128 *Spattering*

1.3.2 CONTAINERS FOR SPRAYING

Paint is delivered to a spray gun by one of three main methods:

Gravity feed
Syphon or suction feed
Pressure feed.

Many variations of these are available, each equipment manufacturer producing their own shape and size of container with its individual characteristics. The choice of which type to buy or use may depend on several factors.

(i) *Type of job:* large jobs require more paint and large containers need filling less frequently.

(ii) *Speed of application:* large flat surface areas can be sprayed quickly with larger systems whereas the smaller units are quite sufficient for small jobs.

(iii) *Type of compressor available:* The amount of air available will govern the type of equipment used.

(iv) *Type of paint used:* some heavy-bodied paints must be applied under pressure, whereas thinner materials can easily be sprayed using low pressures and low-volume units.

Gravity-feed container (Fig. 1.129)
A paint-container or cup of metal or plastic fitted to the top of the spray gun. Paint is fed into the gun as it falls by the force of gravity.
Size Two cup-sizes are usually available holding either $\frac{1}{4}$ or $\frac{1}{2}$ litre of paint.
Use Can be used for most normal types of paint, but not heavy-bodied material such as plastic texture paints.

The small cup-size and limited angle of working limits its use to staining, preparing patterns and small areas (e.g. radiators and touch-up jobs).

Syphon- or suction-feed container
(Fig. 1.130)
The container or cup is screwed underneath the gun. The air nozzle is designed in such a way (see 1.3.1) that a low vacuum is created in front of the fluid nozzle. This allows atmospheric pressure to force the paint up the syphon tube to the fluid nozzle.
Sizes Cups are available holding either $\frac{1}{2}$ or 1 litre of paint (1 litre is the maximum because of weight).

Fig. 1.129 *Gravity-feed container*

Fig. 1.130 *Syphon- or suction-feed container*

Use For spraying small quantities of paint or where several colour changes are required. Not suitable for heavy-bodied paints. Many pressure-feed guns can be adapted to take syphon cups.

Pressure-feed container

Paint is stored in a closed metal container and forced out to the spray gun by compressed air. The amount of paint fed to the gun is controlled by increasing or decreasing the air pressure in the container. Two main types are available:

Cups – small quantities up to 2 litres.

Tanks – various sizes from 10 to 200 litre capacity.

Being under pressure, they have to be tested and approved for safety. Cups are tested from 40 psi (2.7 bar) to 50 psi (3.4 bar) and tanks to 110 psi (7.5 bar).

Pressure-feed cup (Fig. 1.131)

The cup is fitted under the gun as in the syphon-feed container. A small amount of the compressed air going to the bleeder gun is directed into the cup. This forces the paint up the fluid-tube to the fluid nozzle. (Working pressure safe up to approx. 40 psi (2.7 bar).)

Size Usually 1 litre cup.

Use Mainly with a small low-pressure compressor producing approximately 2 cfm (0.95 litre/s), to apply small quantities of most standard decorative paints, but not heavy-bodied materials.

Note A special type of pressure-feed-cup gun is available for higher pressure work and for heavy-bodied paints, used mainly for testing paints in laboratories.

Remote pressure-feed cup (Fig. 1.132)

A small pressure-cup detached from the gun, with all the advantages of the larger pressure-tanks but light enough to be carried in the hand. It has a short hose and is easily controlled as all the controls are in the hands of the operator.

Working pressure up to 50 psi (3.4 bar).

Size 2 litre cup capacity.

Use Ideal for applying most types of decorative paint, including multi-colour finishes. Useful when working from ladders or scaffolding and on smaller jobs where 2 litres is a convenient amount to use.

Pressure-feed tank (top outlet) (Fig. 1.133)

Made from heavy-gauge steel with a lid which is clamped and screwed down. Some types have a separate container which fits into the pot and is removed for cleaning purposes. Most are fitted

Fig. 1.131 *Pressure-feed cup*

Fig. 1.132 *Remote pressure-feed cup*

1 *fluid to gun*
2 *pressure gauge*
3 *air in*
4 *air to gun*
5 *safety valve*
6 *agitator*
7 *fluid-pipe*

Fig. 1.133 *Pressure-feed tank (top outlet)*

with an agitator to allow any heavily pigmented materials which might settle out to be stirred while they are in the tank. Special air-driven agitators are available to allow constant stirring of materials which are likely to settle quickly.

Size From 10 to 200 litre capacity. Tanks holding 25 litres are ideal for larger decorating contractors.

Use Ideal for jobs where large quantities of paint are to be applied. Advantages include:

  (i) Time is saved in refilling.
 (ii) Application rate is fast.
(iii) Paint delivery is constant.
 (iv) Paint cannot be spilt while spraying.
  (v) The gun can be tipped to any angle when spraying.

Pressure-feed tank (bottom outlet) (Fig. 1.134)
Identical in most ways to the top-outlet type except that the outlet is at the bottom of the tank, so that an insert container for easy cleaning cannot be fitted.

Size Available in various sizes from 10 to 80 litre capacity.

Use For heavy-bodied materials such as texture paints, anti-condensation and heavy bituminous compounds which are too thick to be forced up the normal fluid-pipe in top outlet tanks.

1 *pressure gauge*
2 *air in*
3 *air to gun*
4 *safety valve*
5 *follower plate*
6 *texture paint*
7 *fluid out*

Fig. 1.134   *Pressure-feed tank (bottom outlet)*

1.3.3 COMPRESSORS, RECEIVERS, TRANSFORMERS

Compressor
A machine driven by an electric, petrol or diesel engine and designed to drawn in air from the atmosphere and compress it by decreasing its volume and increasing its pressure. A continuous flow of air must be supplied at the required pressure and volume to suit the equipment being used.

Many types and sizes are available. Three main types, available as static or portable units, are used by the decorator.

(i) *Piston compressor* (Figs 1.135 and 1.136)
This is the type of compressor most commonly used. It consists of a piston inside a cylinder, driven by a crankshaft. Air is taken in through filters, compressed by the piston and then stored in an air receiver.

If the required pressure is obtained by a single action, the compressor is described as *single-stage*. This is efficient only up to 100 psi (6.5 bar).

Fig. 1.135   *Piston compressor: single-stage (compressed twice)*

Fig. 1.136   *Piston compressor: two-stage (compressed twice)*

Higher pressures required to be maintained at or above 100 psi (6.5 bar) cannot efficiently be produced by a single action. In this case, air discharged from the first piston is fed into another smaller cylinder and compressed still further – *two-stage* compression.

Where air is compressed there is an increase in temperature, and some form of cooling has to be employed. This is either done by water or, on small units, fins are fitted on the outside of the cylinder.

(ii) *Diaphragm compressor* (Fig. 1.137)
Air is compressed by the up-and-down movement of a diaphragm, rather like the action of a bicycle pump. This type has a relatively small capacity and is used for spray units which produce low pressures only.

(iii) *Rotary compressor* (Fig. 1.138)
Air is compressed by a number of vanes like a fan. As they rotate, the air is gradually compressed.

Static compressors are installed permanently in sprayshops. They are large units, usually bolted to the floor, with air-receivers large enough to supply all the equipment installed in the workshop (Fig. 1.139). They should be placed in such a position as to be able to supply all the air-transformers strategically placed around the workshop.

Portable compressor units are designed so as to be easily transportable. They are usually fitted with wheels or castors, but some of the smaller plants can be carried easily by one person. They vary in size from small units which can be carried in the boot of a car to large compressors which have to be towed by a vehicle (Fig. 1.140).

Air-receiver
The air-receiver is a vertical or horizontal storage-tank which stores the compressed air at high pressure. It must be strong and airtight, and must be cleaned and examined at least every 26 months. A solid-drawn receiver must be checked at least every four years.

Each receiver must be fitted with:
(i) A safety valve to prevent the tank bursting if too much pressure is built up.
(ii) An accurate pressure gauge.
(iii) A tap at the bottom to drain off any oil or water.
(iv) An inlet and an outlet valve.

1 air relief valve
2 upper valve
3 lower valve
4 air outlet
5 diaphragm
6 piston
7 eccentric gear
8 air inlet filter

Fig. 1.137   *Diaphragm compressor*

Fig. 1.138   *Rotary compressor*

1 compressor
2 electric motor
3 air-receiver
4 air outlet
5 air inlet filter

Fig. 1.139   *Static compressor*

1 diesel, electric or petrol engine
2 compressor
3 air-receiver
4 air outlet
5 air inlet filter

Fig. 1.140   *Portable compressor*

Air-transformer (Fig. 1.141)
Apparatus fitted at various points around a spray shop, or on a portable compressor, to which airlines are attached. It strains the air, removing any oil and water in the line, filters any dirt and regulates the amount of air going to the gun.

Fig. 1.141   *Air-transformer*

1.3.4 HOSES AND CONNECTIONS

Air hose (Fig. 1.142)
Air hoses are made of high quality PVC (polyvinyl chloride) reinforced with high tensile braiding. They are made to stand maximum working pressures of 210–230 psi (14.5 to 15.8 bars).
Size   Available in 3 sizes:
6 mm ID (internal diameter)
8 mm ID – most commonly used
11 mm ID.
Pressure drop
When the air pressure at the gun is less than the air pressure at the transformer or compressor, this is known as 'pressure drop'. Although the interior lining of the hose is smooth, it still causes some resistance to the flow of air by friction, resulting in such a drop in air pressure. Therefore (a) the smaller the internal diameter, the higher the drop in air pressure; (b) the longer the air-line, the greater the drop in air pressure.
   Table 1.5 shows how air pressure can drop in various lengths of 6 mm and 8 mm ID hose.

Fig. 1.142   *Air hose*

Fluid-hose (Fig. 1.143)
The internal lining of a fluid-hose must withstand not only fluid pressure but also water, white spirit, oil, cellulose and synthetic paints and their solvents. The hoses are made from compounded rubber or special plastic materials, mainly Thiokol, with a single or double braid reinforcement.
Size   9 mm ID for normal use; 12 mm ID for heavy-bodied paints and where lengths longer than 6 m are used; 19 mm ID special hose for use with bottom-outlet pressure pots to apply heavy-bodied materials such as plastic texture paints.

Fig. 1.143   *Fluid hoses*

TABLE 1.5   **Pressure drop related to hose size**

ID of hose	Air pressure drop at spray gun			
	3 m length	*4.5 m length*	*6 m length*	*15 m length*
6 mm at 40 psi or 2.7 bar	8 psi or 0.5 bar	9.5 psi or 0.6 bar	11 psi or 0.75 bar	24 psi or 1.6 bar
6 mm at 60 psi or 4.1 bar	12.5 psi or 0.8 bar	14.5 psi or 1.0 bar	16.75 psi or 1.1 bar	31 psi or 2.1 bar
6 mm at 80 psi or 5.5 bar	16.5 psi or 1.1 bar	19.5 psi or 1.3 bar	22.5 psi or 1.6 bar	37 psi or 2.5 bar
8 mm at 40 psi or 2.7 bar	2.75 psi or 0.2 bar	3.25 psi or 0.25 bar	3.5 psi or 0.25 bar	8.5 psi or 0.6 bar
8 mm at 60 psi or 4.1 bar	4.5 psi or 0.3 bar	5 psi or 0.35 bar	5.5 psi or 0.4 bar	11.5 psi or 0.8 bar
8 mm at 80 psi or 5.5 bar	6.25 psi or 0.45 bar	7 psi or 0.5 bar	8 psi or 0.55 bar	14.5 psi or 1.0 bar

Note   psi = pounds per square inch; the metric equivalents in bars are approximate.

Hoses for airless spray (Fig. 1.144)
Especially strong high-pressure fluid hoses made with either
(a) A smooth nylon lining reinforced with nylon braid inside a vinyl or nylon cover. This type has a built-in spiral static wire.
(b) A tough plastic (PTFE) lining with a stainless steel braid cover.
Size   Available with 3, 6 or 9 mm ID.

Fig. 1.144   *Hoses for airless spray*

Hose connection
Two types are available:
(i) The more commonly used screw-on type for fluid and air-hose connections (Fig. 1.145). Available in several sizes, they vary according to type of hose and manufacturer.

Fig. 1.145   *Screw-on hose connection*

(ii) Quickly detachable or bayonet type which only requires plugging in: self-sealing and can be used on air or fluid-hoses (Fig. 1.146). These are useful for work which requires frequent changes of gun, or in factories where equipment is run off a long fixed air-supply line. (Not used for airless spray.)

1.3.5 TYPES OF SPRAY SYSTEM

Many types of spray system are available to suit almost any type of work. Different combinations of the various components make the choice even wider. The four main examples shown are typical of systems which can be used by the painter and decorator.

Portable lightweight compressor (Fig. 1.147) A simple unit which can easily be carried in the boot of a car. It is useful for many small decorating jobs including the application of undercoats, gloss and eggshell finishes, emulsion paint and multi-colour finishes.

Fig. 1.146   *Bayonet-type hose connection*

Small 1/2 or 1/3 hp rotary or diaphragm–type compressor electrically driven no receiver or transformer produces approx 2 cfm (0.95 litre/s) pressures of 30–40 psi (2 bar) weight approx 30 kg

bleeder–type low–pressure–feed spray gun requiring approx 2 cfm (0.95 litre/s) at 30–40 psi (2–2.7 bar)

or

bleeder–type spray gun with remote pressure cup requiring approx 2 cfm (0.95 litre/s) at 30–40 psi (2–2.7 bar)

Fig. 1.147   *Spray system: portable lightweight compressor*

1 suction–feed spray gun
requiring 5–7 cfm
(2.4–3.3 litre/s)
at pressures of
30–50 psi (2–3.4 bar)

or

1 pressure–feed gun with
pressure pot for faster production
requiring 5–7 cfm
(2.4–3.3 litre/s)
at pressures of
45–65 psi (3–4.43 bar)

or

1 air-operated orbital sander
requiring 6–9 cfm
(2.8 4.5 litre/s)
at 50–70 psi (3.4–4.7 bar)

1 1/2 – 2 hp petrol, diesel or electrically
driven compressor delivering
approx 7 cfm (3.3 litre/s)
at 60–70 psi (4–4.7 bar)
weight approx 125 kg

Fig. 1.148   *Spray system: small portable compressor*

3 pressure–feed guns with
pressure pots spraying
standard decorative paints,
each gun requiring approx
5 cfm (2.36 litre/s)
at 60–70 psi (4.1–4.7 bar)

or

1 bottom–outlet pressure pot
and heavy duty gun to spray
texture paints, requiring
12–17 cfm (5.6–8.1 litre/s)
at 40–50 psi (2.75–3.4 bar)

or

1 descaling pistol requiring
10 cfm (4.72 litre/s)
at 80–90 psi (5.45–6.1 bar)
also

1 pressure–feed spray gun
and pressure pot

5–7 hp petrol, diesel or electrically
driven compressor delivering
approx 18–20 cfm (8.5–9.45 litre/s)
at 60–70 psi (4–6.8 bar)
according to load
weight approx 375 kg

Fig. 1.149   *Spray system: heavy-duty portable compressor*

Small portable compressor (Fig. 1.148)
Used for medium-size decorative, maintenance and industrial painting. Can be used to apply most types of paint.

Heavy-duty portable compressor (Fig. 1.149)
Used for larger contract work where two guns or powered hand tools are in constant use.

Static compressor (Fig. 1.150)
Large stationary compressor for use in spray-shops. Available in various sizes from 1 to 15 hp motor, delivering up to 50 cfm (23.6 litre/s).

Hot spray (Fig. 1.151)
Thick oil or synthetic paints are difficult to apply. They can be thinned in two ways:
 (i) Add thinner, which must evaporate before the paint can dry and results in a thinner paint film.
(ii) Heat the paint, whereby it becomes thinner, easier to apply, flows out better and dries faster. By incorporating a heating unit into a normal spray-painting system these properties can all be used to advantage.

Equipment
A conventional air-compressor, spray gun, container and hose are used, but with a heater fitted between the container and gun. A specially jacketed fluid-hose keeps the paint warm.

Use Mainly for factory production spraying but can be used on site to spray oil or synthetic paints on large work.

Advantages
 (i) Saving in thinners.
 (ii) Paint can be sprayed at lower air pressures, reducing overspray.
(iii) The equivalent of two coats of paint can be applied in one application.
(iv) The warm paint flows out better and gives a better finish.
 (v) The solvents that are used evaporate quickly and the paint dries faster.

Disadvantages
 (i) Not suitable for water-thinned materials.
(ii) The equipment is costly and not easily manoeuvrable.

Fig. 1.151   *Spray system: hot spray*

Fig. 1.150   *Spray system: static compressor*

Electrostatic hand gun (Fig. 1.152)

A very specialised method of spraying which incorporates the basic law of electricity that 'unlike electrical charges attract each other'.

Paint is sprayed from a pressure-feed tank through a fluid-hose to the gun in the usual way. At the same time, the paint is given a negative electrical charge. As the electrically charged paint leaves the gun, it atomises and each particle is attracted to any earthed object in its path – i.e. the article being sprayed. This causes the paint not only to cover the face of the object but to wrap round behind to get to earth. The result is that some articles, such as tubes, can be coated all round when sprayed from only one side.

Equipment

An air-compressor, paint-container, electrostatic voltage supply and special spray gun.

The voltage supply usually comes from the mains but small mobile battery-run units are available which can be used on open sites.

Use Many industrial applications; the mobile units are especially useful for painting large areas of railing and chain-link fencing.

Advantages

 (i) No overspray. Very clean to use.

(ii) Great saving in paint.

Disadvantages

 (i) Not suitable for water-thinned paints or heavily pigmented metallic coatings.

(ii) The equipment is costly and not easily manoeuvrable.

1.3.6 AIRLESS SPRAY

Airless spray is created by forcing paint at extremely high pressures through an accurately designed small hole or orifice. As the paint leaves the gun and meets the atmosphere it expands rapidly. These two factors cause the paint to break up into an extremely fine, very even spray pattern. No air is used to atomise the paint, hence the description 'airless'.

Equipment

Fluid-pump

This sucks in paint from an unpressurised container and forces it out at very high pressures through a fluid-line to the gun. There are three methods of forcing out the paint:

(i) *Air motor* Compressed air is supplied from an independent air-compressor. Pressures between 60 and 100 psi (4 to 7 bar) are required with a minimum consumption of 3.5 cfm (1.7 litre/s). The pumps have ratios of from 23:1 to 48:1 according to type and make: i.e. the pump multiplies every pound or bar of air taken in by the ratio number. A 30:1 ratio pump with an air-intake pressure of 100 psi (6.8 bar) will produce spraying pressures of 3000 psi (204 bar) – i.e. $30 \times 100 = 3000$. (Fig. 1.153).

Fig. 1.152 *Electrostatic spray equipment*

Fig. 1.153 *Fluid-pump*

(ii) *Electrically driven hydraulic pump* Hydraulic fluid pressure is used to drive the fluid-pump. This is more efficient and requires no air pressure, having only to be plugged into the main electricity supply. Capable of producing spraying pressures of 3000 psi (204 bar).

(iii) *Diaphragm pump* an electrically driven motor operating a diaphragm-type hydraulic pump. It has a low pump-capacity but works at much greater speeds than a normal pump. This produces pressures of up to 3000 psi (204 bar) delivering a surge-free paint flow of up to 3 litres per minute. It has fewer working parts, is smooth-running and quiet to operate, has no friction or heat wear and is therefore easier and cheaper to maintain (Fig. 1.154).

Fluid-line
Special high-pressure, solvent-resistant small bore hose (see 1.3.4).

Fluid-tip Made from tungsten carbide for long life. The orifice size controls the quantity of paint passing through. The angle of the tip controls the size of the fan pattern. Tips have to be changed when adjustments are required. A wide range is available (Table 1.6).

Reversible head As the orifice size of the fluid-tip is so small, it can easily be blocked by small particles of dirt. To save removing, cleaning and replacing the tip each time this occurs, special heads have been designed to allow the tip to be reversed so that the dirt can be blown out and the tip then returned into position. There are several types available including those where the

Fig. 1.154   *Diaphragm pump*

Fig. 1.155   *Airless spray gun*

Gun (Fig. 1.155)
Specially designed, well-balanced gun with a trigger but no other controls. Fan pattern and material flow is controlled by the fluid-tip. A safety catch is fitted to prevent accidental spraying. The needle, tip and seating are made of tungsten carbide to withstand the high pressures and abrasive action of some paints.

TABLE 1.6   **Fluid-tips in common use with decorative paints**

Type of material	Orifice size	Orifice angle	Approximate spray pattern
Lacquers	0.225 mm	40°	180 × 75 mm
	⌈ 0.275 mm	25°	275 × 50 mm
Thin primers	⌊ 0.275 mm	40°	300 × 75 mm
Gloss paint	⌈ As above, also:		
Undercoat		25°	180 × 50 mm
Eggshell	0.325 mm	50°	350 × 50 mm
Emulsion	0.325 mm	40°	275 × 50 mm
Fillers	⌊ 0.375 mm		
Heavy primer surfacers	0.450 mm	65°	375 × 50 mm

Note   In some orifice sizes angles of 5°, 15°, 80° and 95° are also available.

tip is reversed by a single 180% turn of a lever
(Fig. 1.156).

Filtering (Fig. 1.157)

Fine-gauge filters are fitted over the pump paint-
intake and under the gun fluid-feed connection
but these can easily become clogged if
unstrained paint is used. Vacuum-assisted paint-
strainers are designed to strain paint quickly and
efficiently before use. They consist of a large
circular funnel with detachable wire gauze fitted
to a lid. This is fitted over a container. Compressed air is then blown through a special jet
causing a partial vacuum in the container. This
draws the paint through very quickly.

Advantages of airless spray

(i) Airless spray applies most types of paint
much faster than any other manually
operated method of paint application.

(ii) As the paint container is not under pressure, the pump can operate direct from the
manufacturer's can.

(iii) Because no air is used there is no overspray. A degree of paint 'drop-out' occurs
but this can be reduced by pressure
control.

(iv) A uniform thick coating is produced,
reducing the number of coats required.

(v) A very wet coating is applied, ensuring
good adhesion and flow-out.

(vi) Most paints can be sprayed with very little
thinning; with less solvent the paint sets
faster.

(vii) No thin edges on the spray pattern means
no overlapping on each stroke. With wide

tip in spray
position
blocked

tip reversed
to blow out
stoppage

paint flow

Fig. 1.156   *Reversible head*

paint

filter gauze

air in

paint
sucked in by
partial
vacuum

Fig. 1.157   *Vacuum-assisted paint-
strainer*

fan patterns this makes airless spray much
faster than conventional spray.

(viii) Because of the even spray pattern and
clean edge, cutting in at angles can be
achieved.

(ix) The single hose connection to the gun
makes it easy to handle.

1.3.7 CLEANING, MAINTENANCE AND STORAGE OF SPRAY EQUIPMENT

Cleaning

Perfectly clean spray equipment is essential to trouble-free spraying. Dirty and badly maintained equipment is the main cause of breakdown. Remember:

(i) It is much easier to clean paint from equipment immediately after spraying, while it is still wet. Dry hard paint is very difficult to remove.

(ii) Use the solvent recommended for the type of paint that has been sprayed:
Water for emulsion paints
White spirit for oil and oil/resinous paints
Cellulose thinners for cellulose paints and lacquers
Special solvents as supplied for special paints
Never use caustic solutions as they will attack the aluminium used for some parts of the equipment.

Cleaning air spray equipment

Cleaning the paint container
The method will vary slightly according to the type of container used, but the sequence is:

(i) Remove all remaining paint from container.

(ii) Where necessary, blow back to remove paint from the fluid hose.

(iii) Spray several small amounts of clean solvent through the system until it is perfectly clean.

(iv) Wipe the container outside and inside, including lid gaskets, with solvent until perfectly clean.

(v) Ensure that vent holes in gravity and suction feed pots are not blocked.

Cleaning the fluid hose
Provided it is carried out efficiently, the above process will also clean out the fluid hose. It is essential that all paint is removed as (a) paint left inside will slowly build up and reduce the bore, and (b) dried paint will break away and contaminate fresh paint, causing gun blockages.

Hoses should be blown through with air to dry them, or hung over a peg to drain and dry.

Ensure also that the connections are clean and not damaged.

Cleaning the spray gun
Correctly carried out, the above process will also remove paint from inside the gun.

Following this, the air cap should be removed, scrubbed clean, and any paint in the holes removed with a thin stick or fibre (never metal). At the same time make sure that the outside of the fluid-tip and all the outer guncasing is wiped clean.

Never leave the gun to soak in solvent. This destroys the grease around the fluid-needle spring and dries out the valve packings.

Cleaning airless units

(i) Reduce or cut off pressure.

(ii) Remove fluid-tip and clean.

(iii) Remove paint from equipment or pump from paint.

(iv) Drain off paint from fluid-hose.

(v) Place solvent in fluid-container or place pump in solvent-container.

(vi) Set pump at low pressure and circulate solvent until system is clean (Fig. 1.158).

(vii) Wipe all units clean ready for storage.

Fig. 1.158 *Cleaning the airless spray unit*

Maintenance

When the equipment is clean and dry the following simple maintenance operations should be carried out.

(i) The guns should be lubricated and greased to reduce wear and ensure free movement of all working parts (Fig. 1.159).

(ii) Coil up air and fluid hoses without twists or kinks.

(iii) Open the drain cocks on the receivers and air transformers to let out any moisture and oil droplets. Leave open until used again.

(iv) Electric cables should be coiled without kinks or twists.

Storage

Compressors and plants should be kept in a dry store where they cannot be knocked or damaged by exposure to weather.

Hoses should be coiled, tied and hung over two pegs in a dry store.

Guns are safest hung in a cupboard or laid in a drawer in a warm dry store.

Oil: light machine oil
Grease: petroleum jelly

Fig. 1.159 *Lubrication of spray equipment*

1.3.8 SPRAY GUNS FOR SPECIAL PURPOSES

Air brushes (Fig. 1.160)

A range of small, low pressure spray guns, capable of very fine adjustments to the air and fluid controls. This enables a spray pattern to be produced ranging from a fine hair-line, to a broad even spray.

Usually fitted with an internal mix air cap. A spatter effect can also be produced with a special air cap. Air brushes are fitted with either a gravity feed, or a suction feed fluid container. The fluid containers are small, and range from 5 ml to 34 ml in capacity.

The guns operate on compressed air supplied by either small portable air compressors containing an air receiver, or disposable cans of aerosol propellant.

Use Vehicle customising, edge sprayed motifs, spray stencilling, decorative shading and spattering.

Fig. 1.160 *Air brush*

Hopper feed gun (Fig. 1.161)

A range of gravity feed spray guns specially designed for the application of heavy texture coatings.

The guns operate by allowing heavy texture material to fall by gravity into a stream of compressed air. The air pushes the material through a small aperture in a nozzle, and expands it into a wide, but coarse spray.

The material is carried in a lightweight plastic hopper fixed above the gun. Hoppers have a capacity of approximately 4.5 litres (1 gallon). The spray nozzles have apertures ranging from 5 mm to 7 mm diameter. Also available is an elbow adaptor for spraying floors and ceilings.

The guns consume about 200 litres/second (7 cfm), at a pressure of 2.75 bars (40 psi), and can be operated with small portable air compressors containing an air receiver, and delivering a minimum of 230 litres/second (8 cfm), at a pressure of 3 bars (45 psi).

Use For the rapid and economical application of texture coatings such as cement paints, texture relief coatings, non-slip deck paint, stone paints, masonry paints, emulsion paints.

Fig. 1.161 *Hopper feed gun*

Flocking gun (Fig. 1.162)

A range of spray guns specially designed for the application of flock to a surface which has been previously coated with a suitable adhesive. The flock adheres to the surface and produces a soft textured finish similar to felt.

The guns operate by directing a low pressure stream of compressed air into a container of flock attached beneath the gun. The stream of air disturbs the flock, and blows it through a simple nozzle, which directs the flock onto the sticky adhesive.

The guns can be operated with any air compressor containing an air receiver.

Flock Cotton or nylon fibres available in a range of colours. The fibres are very fine (2 denier), and approximately 2 mm long.

Use Decorative motifs, stencils, vehicle customising.

Fig. 1.162 *Flocking gun*

1.4 POWER SOURCES

1.4.1 COMPRESSED AIR

Air can be used to drive certain types of air-operated mechanical hand-tool and is used in spray painting. To be useful, it must have pressure above normal atmospheric pressure (approx 15 psi or 1 bar). Air pressure is increased by compressing the air into a smaller volume: the more a given quantity of air is compressed, the greater the pressure.

Example (Fig. 1.163)
1 cubic metre of air at 15 psi or 1 bar = $\frac{1}{2}$ cubic metre of air at 30 psi or 2 bar = $\frac{1}{4}$ cubic metre of air at 60 psi or 4 bar.

Air is compressed and raised to the required pressure by mechanical means in compressors.

When choosing air compressors and spray equipment it is important to understand the difference between air volume, air pressure, air displacement and air delivery.

Air volume is the amount in cubic feet or cubic metres of air at any pressure, i.e. the volume of air in the above example is 1 cubic metre, $\frac{1}{2}$ cubic metre and $\frac{1}{4}$ cubic metre. (The bigger the orifice of a spray gun the more air it lets out and the greater the volume of air it consumes.)

Air pressure is the pressure in pounds per square inch (psi) or bars (1 bar = approx 15 psi)

Fig. 1.163 *Compression of air*

TABLE 1.7 **Air delivery/air pressure requirement of equipment**

Equipment	Volume of air required (air delivery)	Air pressure required
Spraygun (depending on type and size)	2–10 cfm or 0.944–4.720 litre/s	Average 50–70 psi or 3.3–4.6 bar
Descaling pistol	5–10 cfm or 2.360–4.720 litre/s	100 psi or 6.5 bar
Large abrasive blasting equipment	330 cfm or 155 litre/s	100 psi or 6.5 bar

exerted by air which has been compressed. The greater the pressure, the greater the force: it is the air pressure that atomises the paint when it is sprayed.

Air displacement is the amount of air that can theoretically be produced by a compressor in a given period of time. It is measured in either (a) cubic feet per minute (cfm) or (b) litres per second (litre/s).

Air delivery is the actual amount of air delivered to the gun or receiver. No compressor is 100 per cent efficient (usually 60–80 per cent). Therefore, although an air displacement figure is quoted for each compressor, this does not represent the amount of air it can actually deliver to the air receiver or the gun. For instance, if an air compressor has a displacement of 10 cfm (4,720 litre/s) and it is only 70 per cent efficient, the delivery or amount of air actually available for use will be only 7 cfm (3.304 litre/s).

The above definitions are important as, to run each gun or piece of equipment, a specific volume of air within a given time is required: for maximum efficiency, a specific air pressure is also required (see Table 1.7).

Atomisation is the breaking up of a liquid into small droplets, usually in a high-speed jet of air or by forcing it through a very small orifice at high pressure, as in a hose pipe (Fig. 1.164).

Kingdom are designed for 240 V, or 110 V for equipment used on construction sites. The required voltage is usually stated on the appliance.

Ampere or amp (A) This is the measure of the *quantity* of electricity which flows along the conductor or wire. Fuses, cables and plugs are all rated according to the quantity of electricity they can carry. To calculate amps, divide wattage by voltage. *Watt (W)* The watt is the unit of the *power* of electricity; all appliances state how much power is required to operate them, e.g. a 100 W bulb. Wattage can be calculated by multiplying volts by amps. As a watt is a relatively small unit, the kilowatt is often used:

1000 W = 1 kW.

Example

Most modern houses have a 13 amp ring main system with a voltage input of 240 volts, i.e. $13 \times 240 = 3120$ W. If a number of appliances with a total wattage above 3120 watts are all switched on together, the system will be overloaded and the fuses will blow.

AC/DC

All main supplies are alternating current (AC) whereas local generators are direct current (DC). All power tools are supplied for AC, some for both AC and DC.

1.4.2 ELECTRICITY

Electricity is composed of small particles, known as electrons, which flow through conductors such as wires. It flows only when it can complete a circuit, i.e. from the source, along a wire and back to the source. Domestic electric cables contain at least two wires which are colour-coded: electricity flows from the source by way of the *live* brown-sheathed wire to the appliance and back along the *neutral* blue-sheathed wire (Fig. 1.165 a). If the wires are joined or in some way are able to touch each other, a short circuit occurs; this blows the safety fuse (Fig. 1.165 b).

In power stations electricity is produced at very high voltages; these are reduced by a series of transformers to 240 volts for domestic use.

Units of measurement

Volt (V) Electricity flows along a conductor as a result of electrical *pressure* which is measured in volts. Most electrical appliances in the United

Fig. 1.164   *Atomisation*

Fig. 1.165   *Electrical circuit*

Transformer

A portable apparatus for increasing or reducing voltage: its main use is to reduce 240 V mains supply to permit the use of 110 V appliances.

Insulators and conductors

A conductor is a material which allows electricity to pass through it. Most metals are good conductors of electricity, the best being copper and silver.

An insulator is a material which will not allow electricity to pass through it. Most non-metallic materials are good insulators, the best being rubber, glass, ceramics and plastics.

Single phase and three phase electricity

All power hand tools, e.g. drills, sanders, angle grinders, are powered by single phase electric motors. Single phase motors operate from either the normal domestic electrical installation of 220/240 V, or this domestic voltage transformed to a safer level of 110 V.

For producing higher levels of power than that required by hand tools, three phase motors are used. Factory machinery and some air compressors use three phase motors. Special electrical installations are necessary to operate these motors, and they cannot be connected to the normal domestic supply. The connecting plug contains at least five pins and the voltage measured across these pins can be in excess of 400 V.

Because three phase supply is rare on most sites on which a painter may work, a portable three phase electric air compressor is of little value. They are ideal for static equipment in a properly installed workshop where their greater power and efficiency can be used to their best advantage.

Flex and cables

The word 'flex' is an abbreviation of 'flexible cord'. Flex is used to connect electrical equipment to a power point. Each wire or conductor is made up of a number of thin strands, which gives it flexibility.

Cable is used for mains wiring. It is more rigid than flex, the conductor being a single thick wire strand.

Fuses

There is a limit to the amount of electricity that can safely pass through a wire. If the system is loaded above this limit, heat develops, causing the insulation to break down and a short circuit to occur. To prevent this happening, fuses are fitted into the circuit or appliance: these are weak-spots in the system, designed to fail as soon as the circuit is overloaded. It is therefore vital that the correct fuse should be fitted since, if a higher grade fuse or piece of metal is used, it may allow too much overheating resulting in damage to the appliance or even in fire.

There are two types of fuse:
(i) The rewirable type where various grades of wire can be used according to the quantity of electricity being used in the circuit.
(ii) The cartridge type, where the required grade of wire is enclosed in a small cartridge (Fig. 1.166 a). This is used in all fused plugs, some appliances and switch fuses.

Earthing

All exposed metalwork on meters and appliances must be earthed. This ensures that if any appliance becomes faulty and the metal framework becomes alive, the current will flow to earth, blow the fuse and prevent any risk of electric shock.

Only special appliances which are completely insulated from the user need not be earthed. These are marked 'double insulated' (Fig. 1.166 b).

Residual current circuit breaker (RCCB)

A small piece of apparatus connected to a socket outlet or fuse box which immediately cuts off the electrical power if a fault occurs, such as a cut flex, so protecting the user from a possibly fatal electric shock.

(a)

(b)

a *Cartridge fuse*
b *'Double insulated' symbol*

Fig. 1.166

The international wiring code (Fig. 1.167)

E or ⏚ is earth and is shown by green and yellow bands.

L is live and is coloured brown.

N is neutral and is coloured blue.

Electric shocks

An electric shock occurs when the body becomes part of an electric circuit. This can happen when the body comes between either (a) neutral and earth or (b) live and earth.

The resistance of a human body to electric shock varies according to the condition of the body, age, situation and the weather. A shock of 40 V may be dangerous, and 240 V can cause death or severe burning. Because of this, most new building sites in this country have the voltage reduced to 110 V during construction. This greatly reduces the chance of a fatal accident.

Special equipment is necessary for 110 V: 240 V equipment *is not suitable* unless a transformer is used.

For treatment for electric shock see 6.8.2.

Safety precautions

Electricity is an unseen hazard. All electrical gear is a potential source of danger.

Never   use too many appliances on one circuit: overloading is dangerous.

Never   tape joints: use junction boxes or insulated connectors.

Never   drag cables through pools of water.

Never   fix wires into sockets with matchsticks or pieces of wood.

Never   remove plugs from sockets by pulling the lead.

Never   coil cables incorrectly: they may kink and break.

Never   drag tools by their cable: their connections may break.

Never   use a machine or piece of equipment for a purpose for which it is not designed.

Always wire up correctly: use the correct socket and plug for the appliance or get them wired up by a competent electrician.

Always ensure that the use and control of the equipment is understood before use.

Always check the voltage marked on the equipment and the voltage on the supply. They must be the same.

Fig. 1.167   *Wiring*

Always check for loose or missing screws and for damaged leads.

Always attach earth wire when wiring up.

Always have tools regularly checked for wear and have damaged or worn parts replaced.

Always report defects or damage to the foreman or supervisor.

Note The Construction (Health, Safety and Welfare) Regulations 1996 state that any electrically charged overhead cable or apparatus must be made safe or sealed off. This is to prevent accidents when carrying or moving equipment.

1.4.3 GASES

Two main types of liquefied petroleum gas (LPG) are used in the building industry: butane and propane. They are produced by the distillation of crude oil. The gases are purified, have no odour and are non-toxic. To comply with international regulations on safety of gases, an agent is added to produce a smell which can easily be detected. Both gases are heavier than air and are available in bottles and cylinders containing from 1 to 82 1b (0.45–36 kg).

At normal air temperatures propane and butane are gases, but they will liquefy if cooled and will remain liquid at normal temperatures if kept under pressure in cylinders. Once the pressure is released they vaporise and revert to gas. Blow-torches used by painters require approx 10 psi (0.714 bar). This is controlled by using a pressure regulator fixed to the cylinder.

Gas level

The only way to check the amount of gas remaining in a container is to weigh the cylinder

TABLE 1.8 **Properties and uses of butane and propane gases**

	Butane	*Propane*
Ratio gas : liquid (in volume)	approximately 240 : 1	approximately 275 : 1
Amount of gas from 0.45 kg liquid	6.7 ft^3 (0.190 m^3)	8.65 ft^3 (0.245 m^3)
Temperature at which liquid vaporises	0°C Cannot be used below freezing point	−44°C Usable in all conditions
Thread on cylinder fittings	Left-hand (female)	Left-hand (male)
Vapour pressure	At 15°C = 25 psi At 45°C = 85 psi	At 15°C = 100 psi At 45°C = 285 psi
Use	Mainly domestic, also used by painters for burning-off and with steam strippers	Mainly in industry for heating and lighting

and subtract the tare weight, which is always marked on the outside. (Tare weight is the weight of the empty container.)

Safety precautions

Never look for leaks with a naked flame: use soapy water.

Never repair faulty hoses: replace them.

Always use special reinforced hoses as specified by the manufacturer.

Always keep cylinders away from heat: heated cylinders may explode.

Always keep gas cylinders upright: liquid gas may get through the regulator, causing high pressure at the torch.

Always use a pressure regulator to control pressure and prevent blow-back.

Always ensure plenty of ventilation when using gases indoors.

Other gases

Acetylene is combustible gas and has an unmistakable smell. It is supplied in maroon cylinders with left-hand thread fittings. Only red hoses should be used.

Oxygen is a non-combustible gas but is essential for burning acetylene gas. It has no smell and is supplied in black cylinders. Fittings have a right-hand thread and hoses are black or blue.

Never allow oil or grease to come into contact with oxygen as the two form a mixture liable to spontaneous combustion.

Use Oxygen/acetylene gases are mixed to produce the intense heat required for flame-cleaning.

For *storage* of gases see 6.2 and 6.3.

1.4.4 THE INTERNAL COMBUSTION ENGINE

In an internal combustion engine, a mixture of fuel and air is burned in such a way that the hot gases produced from the combustion exert a force on working parts of the engine, thus generating power.

Internal combustion engines can be divided into two main groups:

(i) **Petrol engines**, in which a mixture of petrol and air is compressed and ignited by means of a spark from a sparking plug. The hot gases produced exert a force and drive the engine.

(ii) **Diesel engines**, where air is compressed into a small space in the top of a cylinder. This creates a very high temperature, hot enough to ignite a low-grade fuel (diesel oil) when it is injected into the cylinder, without a spark.

The cycle of operation of petrol and diesel engines is similar, the only difference being that diesel engines have no sparking plug and run on a lower-grade fuel. Both work on either a 2-stroke or 4-stroke cycle (Figs. 1.168–1.170).

2-stroke engines (Fig. 1.168)
The cycle is completed in a one-up and one-down stroke of the piston, or one turn of the crankshaft.
(i) Compression: in the first stroke the piston rises and compresses the fuel in the top of the cylinder while at the same time fresh fuel is taken in under the piston. At the top of the stroke the mixture is fired by a sparking plug and an explosion occurs.
(ii) Power: as the piston is forced down, it drives the crankshaft; as the piston drops, the exhaust fumes escape, and fresh fuel is forced into the top of the cylinder. The cycle is then complete.

4-stroke engine (Fig. 1.169)
The cycle is completed in two upward and two downward strokes of the piston and two full turns of the crankshaft.
(i) Intake: the piston goes down and a petrol/air mixture is let in through the inlet valve.
(ii) Compression: the piston then rises and the mixture is compressed in the top of the cylinder.
(iii) Power: the mixture is ignited, an explosion occurs and the piston is forced down – 'the power stroke'.
(iv) Exhaust: the piston rises again, forcing all the burnt gases out through the exhaust valve which opens with the fourth upward stroke. The cycle is then complete.

Simple maintenance checks
(i) Always make sure there is fuel in the tank. If it is allowed to become empty, air and dirt may enter the pipes and make restarting difficult.
(ii) Use the correct fuel, i.e. petrol, diesel oil or petrol/oil mixture for 2-stroke petrol engines.
(iii) All oil levels must be regularly checked and topped up to prevent the engine seizing up.
(iv) Where applicable, always check the water level in the cooling system – overheating can cause the engine to seize up.
(v) Remember, regular servicing and correct use prevents breakdown, reduces running costs and increases the life of any engine.

Fig. 1.168   *2-stroke engine*

Fig. 1.169   *4-stroke engine*

Safety precautions

  (i) All exhaust fumes must be piped into the open air as they contain a poisonous gas (carbon monoxide) which can cause headaches, drowsiness and death if inhaled in large enough doses (see also 6.6).

 (ii) Always keep the top on the fuel tank; fuel spilled on to a hot engine may cause fire or an explosion.

(iii) Always make sure that portable engines are on firm ground and wedged to prevent them running away with the vibration, when they may be damaged, causing injury, fire and explosion.

valve assembly

air intake filter

carburettor

fuel intake

spark plug

crankshaft

piston

flywheel

drive to compressor

Fig. 1.170 *Parts of the internal combustion engine*

PART TWO

Surface Coatings and Specialist Materials

2.1 Introduction   67

2.1.1   Composition   67
2.1.2   Pigments and extenders   69
2.1.3   Film-formers   75
2.1.4   Thinners and driers   75
2.1.5   Properties of paint   77

2.2 Primers, undercoats and stoppers   79

2.2.1   Acrylic primer/undercoat   79
2.2.2   Alkali-resisting primer   79
2.2.3   Aluminium wood-primer   80
2.2.4   Calcium plumbate primer   80
2.2.5   Etch primer   81
2.2.6   Lead-based wood-primer   82
2.2.7   Lead-free wood-primer   82
2.2.8   Spirit-based aluminium sealer   82
2.2.9   Penetrating oily sealer or
       stabilising solution   83
2.2.10 Red lead primer   83
2.2.11 Zinc chromate metal-primer   84
2.2.12 Zinc phosphate metal-primer   84
2.2.13 Zinc-rich primer   84
2.2.14 Oil/resin undercoat   85

2.3 Finishing paints   86

2.3.1   Bronze/aluminium paint   86
2.3.2   Eggshell or semi-gloss finish   86
2.3.3   Matt emulsion paint   86
2.3.4   Vinyl silk emulsion   87

2.3.5   Gloss finish   87
2.3.6   Multi-colour finish   88
2.3.7   Masonry paint   89
2.3.8   Cellulose coating   90
2.3.9   Moisture vapour permeable (MVP)
       or microporous coatings   90
2.3.10 Low odour eggshell finish   91

2.4 Special paints   92

2.4.1   Anti-condensation paint   92
2.4.2   Anti-fungus paint   92
2.4.3   Bituminous paint   93
2.4.4   Cement paint   93
2.4.5   Chlorinated rubber paint   94
2.4.6   Damp-proofing compound   94
2.4.7   Epoxy resin paint   95
2.4.8   Flame-retardant paint   95
2.4.9   Micaceous iron oxide paint   96
2.4.10 Fluorescent paint   96
2.4.11 Polyurethane paint   97
2.4.12 Texture paint (exterior)   97
2.4.13 Relief texture paint (interior)   98
2.4.14 Self-cleaning paint   99
2.4.15 Heat-resistant paint   99

**2.5 Clear finishes and wood
     treatments**   101

2.5.1   Wood preservatives   101
2.5.2   MVP stain/varnish   101
2.5.3   Wood stain   102

2.5.4   Emulsion varnish   102
2.5.5   Oil-resin varnish   103
2.5.6   Polyurethane varnish   103
2.5.7   Water-repellent solution   104
2.5.8   Spirit varnish   104
2.5.9   Graining and marbling materials   105

2.6 Paint defects   107

2.6.1   Defects of surface coatings   107
2.6.2   Defects in the tin   111

**2.7 Gilding and bronzing materials
and defects**   112

2.7.1   Gilding   112
2.7.2   Bronzing   113
2.7.3   Gilding defects   115

2.8 Stencil and masking materials   116

2.8.1   Stencil-plate materials   116
2.8.2   Other stencil materials   117

2.1 INTRODUCTION

2.1.1 COMPOSITION (Fig. 2.1)

All surface coatings have two principal components:

(i) A *film-former*, also known as medium, vehicle or binder, whose purpose is
(a) to convert from the liquid coating to a solid dry film;
(b) to bind the pigment particles together;
(c) to provide gloss to the coating;
(d) to make the coating adhere to the surface;
(e) to give elasticity to the film;
(f) to provide resistance to water, chemicals and abrasion;
(g) to hold the pigment in suspension.

The film-formers most commonly used are:

Drying oils – linseed oil; soya bean oil; tung or Chinese wood oil; dehydrated castor oil; oiticica oil.

Resins – copal, rubber, lac, alkyd, coumarone, phenolic, epoxy, polyurethane, polyvinyl acetate (PVA), acrylic, silicone.

Oil-modified resins – natural or synthetic resins mixed with oils, i.e. coumarone resin/tung oil, alkyd resin/linseed oil.

(ii) A *thinner* or solvent (sometimes called a diluent), whose purpose is to make the coating liquid enough to be easily and evenly applied by any method.

It evaporates completely once the coating has been applied.

The thinners most commonly used are water, white spirit, solvent naphtha, xylol, methylated spirit and mixtures of special solvents.

The combination of a film-former and a thinner produces a transparent coating which may be called a varnish, lacquer or clear finish. This type of coating has a limited use.

The most common type of coating in use is *paint*, which is an opaque coloured material. This quality is provided by a third component:

(iii) *Pigment*, according to type, may perform one or more of the following:
(a) to provide opacity or covering-power to the paint;

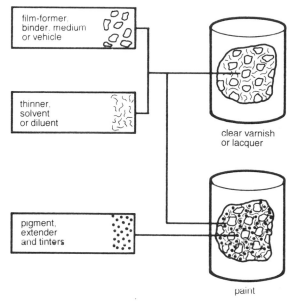

Fig. 2.1 *Composition of surface coatings*

(b) to provide colour;
(c) to aid the film-former in protecting the surface.

The most commonly used pigment is titanium white together with a wide range of coloured pigments and dyestuffs.

Modern surface coatings are complex materials which contain also one or more of the following components, depending upon the type of film-former used or the use to which the coating is put.

(iv) *Drier, hardener or catalyst* ensures that the coating converts from the liquid to the solid state: driers are added to paints containing drying oils to speed up the drying process; curing agents, catalysts or hardeners are used to convert epoxy and polyurethane resins into hard films.

(v) *Plasticiser* is required by some coatings to prevent the film being too brittle.

(vi) *Stabiliser* is required by some coatings to ensure that the complex film-former remains intact.

(vii) *Anti-skinning agents* are added to some oil-based paints to prevent skinning in the tin.

(viii) *Extenders* have various uses in most paints:
(a) they add to the bulk of the paint;
(b) they provide a key in undercoat films for finishing paints to grip;
(c) they prevent pigments settling to the bottom of the paint;
(d) they make some paints easier to apply;
(e) they lower the cost.

(ix) A *thixotropic agent* gives a jelly-like structure to the paint, providing anti-sag and high build properties.

Types of coating

Most coatings used by the painter and decorator can be classified as either (a) *water-borne* – those whose film-former is either soluble in water, or is of a nature that allows it to be suspended in water, or (b) *spirit-borne* – those whose film-former is soluble in a material other than water.

Drying (Fig. 2.2)

The process that converts the liquid material into the solid film is also complex. Most coatings undergo one or a combination of any of the following four processes:

Evaporation   This occurs soon after the coating is applied and is the changing of the thinner into a vapour which is absorbed into the atmosphere. Every material goes through this process, which is recognised by the paint *setting up*, so that it can no longer be brushed or rolled and its flowing property has gone. A few coatings dry solely by this process.

Oxidation   When drying oils are included in the film-former they need to combine with oxygen to complete their chemical change from liquid to solid. Once this has been completed the original film-former has altered its structure and the process cannot be reversed.

Polymerisation   Certain elements in the film-former are combined to produce a solid state: this is a reasonably fast process when applied to spirit-borne coatings which contain no drying oil, but is very slow after initial oxidation has occurred in oil-based paints.

Coalescence   A thinner may have a secondary function, that of keeping apart particles of resin which, if they came into contact, would form a solid film. In this drying process the coalescing, or coming-together, occurs as soon as the thinner has evaporated.

Drying conditions

All coatings used by the painter and decorator are classified as air-drying, meaning that exposure to the atmosphere is essential to start the drying process. Both air and light are necessary to ensure a satisfactory drying action. An absence or shortage of either during the drying process may prevent or retard the forming of a surface coating film. Temperature and moisture also affect the speed of drying.

Types of film

(a) *Convertible coating (non-reversible film)* A coating which goes through a chemical change upon drying, so producing a material chemically different from its liquid state. It cannot be converted back to its liquid state by the use of its original thinner.

(b) *Non-convertible coating (reversible film)* Some coatings are formed by dissolving a resin in a solvent. When the material is applied the solvent evaporates, leaving the resin in its original form spread evenly over the surface. The change has been physical only and the film can be reversed to the liquid state by the application of its original solvent.

Fig. 2.2   *Drying processes*

Types of finish

When surface coatings dry they produce films with varying grades of sheen. The range extends from flat or matt finishes, which have no sheen, through increasing degrees of lustre to high gloss finishes. The degree of sheen is dependent upon the angle at which light is reflected from the surface. A smooth surface gives a uniform angle of reflectance and appears glossy. A rough surface causes a scattering of reflected light waves and appears less glossy or even matt.

Smooth surfaces are produced by finely ground pigments in small proportion to film-former. Rough surfaces are produced by slightly larger pigment particles used in a higher proportion to film-former (Fig. 2.3).

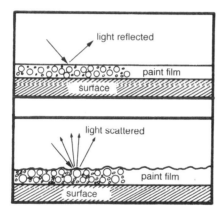

Fig. 2.3 *Types of finish*

Alkyd gloss, matt and eggshell finishes contain similar materials, their difference lying in the sheen resulting from the type, size and quantity of pigment used. Matt paints have a high proportion of pigment to film-former, whereas gloss paints have less pigment and more film-former. This is known as the pigment : binder ratio.

2.1.2 PIGMENTS AND EXTENDERS

Pigment

Pigment is a solid component of paint. A vast number of pigments are used in paint manu-facture, obtained from numerous sources which can broadly be classified as organic and inorganic.

Organic matter is that which at some time has been living. The principal sources of organic pigment are crude petroleum and coal-tar oil. The pigments are produced as dyestuffs (some-times referred to as lakes) in a wide range of colours; because of their fineness, many are struck on to bases of extenders to give them substance. Vandyke brown is one organic pig-ment which is not a dyestuff; it is obtained direct from peaty earth.

Inorganic pigments may be derived from:
(a) coloured earths, sometimes known as earth or natural pigments, which are mixed and ground to a fine powder, e.g. yellow ochre, raw umber;
(b) metals chemically treated to render them into fine powders, e.g. titanium white, white lead, iron oxide;
(c) combinations or heat treatments of chemical compounds, e.g. prussian blue, zinc chromate, lithopone.

Use
 (i) To give body and covering-power to paint: most widely used for this purpose is tita-nium white.
(ii) To colour the paint: a property which all pigments possess.
(iii) To aid the film-former in protecting the substrate; those which do this include red lead, zinc chromate, calcium plumbate.

Stainers
The painter and decorator usually obtains pig-ments in one of the following forms for use in matching colours or staining glazes.
(a) Oil stainers, which are ground in an oil medium to a thick paste and available in pack or tube form.
(b) Universal stainers, which are generally thinner than oil stainers but have the advantage of being soluble in any type of paint.

Classification

Table 2.1 classifies the most common pigments according to colour. Beside each pigment is the following information:
Column 1 indicates its source.
Column 2 states its specific gravity (SG). This is a measure of the pigment's density compared with an equal volume of water: water's SG is 1. Pigments with an SG of 2 are not very dense;

TABLE 2.1 Properties of pigments

Pigment	Earth	Origin Organic	Inorganic	Specific gravity	Opacity	Tinting strength	Resistance to Alkali	Acid	Light	Other characteristics	Principal use
Whites											
Titanium dioxide (rutile)			/	4.2	1	1	1	1	1	Completely stable	In most paints
Zinc oxide			/	5.6	3	3	1	4	1	Becomes brittle	Fungicidal paints
Lithopone			/	4.3	2	2	1	4	2	Chalks on exposure to weather	Undercoats
White lead			/	6.7	3	3	1	5	1	Discoloured by sulphides Toxic	Exterior wood primers
Antimony oxide			/	5.7	2	3	1	4	1	Discoloured by hydrogen sulphide	Fire-retardant paints
Calcium plumbate			/	5.7	2	3	1	4	1	Rust-inhibitive Toxic	Primers for wood and galvanised iron
Zinc phosphate			/	3.3	3		3	3	1	Non-toxic, rust-inhibitive	Metal and some types of universal primer
Yellows											
Yellow lead chromes			/	5.6	3	3	5	4	3–5	Unsuitable for water paints Toxic	Tinter for oil paints
Zinc chromate			/	3.4	4	4	3	3	1	Rust-inhibitive	Metal primers
Ochre	/			2.9	2	3	1	2	1	Tends to retard drying	Tinter for all paints
Iron oxide yellow			/	5.1	2	3	1	2	1	Excellent fastness on weathering	Ideal for finishing paints
Raw sienna	/			4.0	4	5	1	1	2	Transparent in oil	Scumbles and glazes
Hansa yellow IOG		/		1.6	3	3	2	2	1	Excellent fastness on exposure	Decorative paints only
Reds											
Orange and scarlet chromes			/	6.1	3	3	4	4	3–5	Unsuitable for water paints	Tinter for undercoat and gloss
Toluidine red		/		1.4	2	2	1	3	2	Affected by heat	Decorative finishes
Red iron oxide	/			5.1	3	3	1	3	1	Stable	Finishing paints
Red lead			/	9.0	2	3	3	5	4	Rust-inhibitive Toxic	Metal and timber primers
Para red		/		1.5	2	2	2	2	3	Reduced strengths fade in water	Decorative finishes only
Bordeaux		/		1.3	2	2	1	1	2	Unsuitable for cellulose paints	Tinter for most paints
Dioxazine violet		/		1.4	2	1	1	1	1	Expensive	Tinter for all finishes
Blues											
Ultramarine blue			/	2.3	4	3	3	4	3	Loses gloss on weathering	Tinter for all paints
Prussian blue			/	1.8	5	1	5	3	3	Fades in absence of light	Tinter for oil/alkyd paints
Phthalocyanine blue			/	1.6	1	1	1	1	1	Completely stable	All types of paint
Greens											
Phthalocyanine green			/	2.2	1	1	1	1	1	Completely stable	All types of paint
Mid chrome green			/	4.2	3	3	5	4	4	Turns blue on exposure	All types of oil/alkyd finishes
Chromium oxide green			/	5.1	4	4	1	2	1	Heat-resistant	Chemical and heat-resistant paints
Browns											
Raw umber	/			4.0	2	3	1	2	1	Semi-transparent	Scumbles, glazes and tinters
Burnt umber	/			3.5	2	3	1	2	1	Completely stable	Tinter for all paints
Vandyke brown	/		/	1.5	5	4	1	1	1	Brittle and transparent in oil	Water-colour graining
Burnt sienna	/			3.5	3	4	1	1	1	Completely stable	Tinter for all paints
Blacks											
Carbon black			/	1.8		2	1	1	1	Brown undertone	Black gloss finishes
Lamp black			/	1.8		3	1	1	1	Blue undertone	Tinters for all paints
Graphite			/	2.2		5	1	1	1	Leafing properties	Anti-corrosive paints
Metallics											
Micaceous iron oxide	/				1		1	1	1	Leafing properties	Paints for primed steel work
Aluminium powder			/	2.6					1	Leafing or non-leafing types	Sealers and heat-resisting paints
Bronze powder			/	8.9				4	1	Tarnishes on exposure	Decorative metallic paints
Zinc dust			/	7.0	1	4			1	Protects iron and steel	Zinc-rich paints for metal
Metallic lead			/	11.4				1		Rust-inhibitive Settles on storage	Metal primers

TABLE 2.2 **Properties of extenders**

| Extender | Origin | | | Specific gravity | Principal use |
	Earth	Organic	Inorganic		
Barytes	/			4.5	Pigment in water paints Extender mainly in undercoats
Blanc fixe			/	4.0	Improves flow of eggshell finishes
Terra alba	/			2.3	Water paints Primers Sealers Fillers
Witherite	/			4.3	Improves durability and colour permanence
Paris white or whiting	/			2.7	Pigment in cheap water paints Gives body to undercoats
Silica	/			1.9	Good flatting agent Improves inter-coat adhesion Road line paints
China clay	/			2.6	Good opacity in water paints Suspending and flatting agent in oil paints
Fuller's earth	/			2.5	Anti-settling agent
Talc	/			2.8	Flatting, anti-settling and fire-resisting properties
Asbestine	/			2.9	Suspending and flatting agent Used in wood fillers
Mica	/			2.8	Resistant to sinkage, chalking and cracking Good moisture-barrier
Wollastonite	/			2.9	Extender for decorative paints

pigments with an SG of 8 are very dense and would cause paints to be heavy, and would also tend to settle out of the paint in storage.

Column 3 gives a guide to the pigment's ability to obliterate a surface. Those with a rating of 1 have very good opacity, and those with 5 have poor opacity.

Column 4 gives an indication of the power of a pigment to stain a white paint, or of a white pigment to resist staining: 1 = excellent staining power or reducing power; 5 = poor.

Column 5 indicates the pigment's ability to resist the three most common elements which may affect the colour of pigments: 1 = very good resistance; 5 = no resistance.

Extender (Table 2.2)

Extenders are also solid constituents of paint and are used in many types of surface coating. They are obtained from natural materials which are mined and ground to a fine powder. They are generally white in colour, and are transparent in oil, therefore having no influence on the colour of oil paints. They are sometimes referred to as 'mineral whites'.

Properties
 (i) Allow certain paints to be applied with ease.
 (ii) Give greater adhesive properties to undercoats.
(iii) Provide a 'roughness' in the film of undercoats to improve adhesion of the finishing paint.
 (iv) Prevent certain dense pigments settling out of the paint.
 (v) Give bulk to certain paints which do not contain sufficient pigment content, e.g. black, or bright colours which do not contain white.
 (vi) Provide added hardness to paint films.
(vii) Improve the flowing properties of certain paints.

TABLE 2.3 **Properties of drying oils**

Oil	Properties	Use
LINSEED OIL (obtained from the seed of the flax plant)	Pale yellow in colour Dries within 7 days under good drying conditions Saponified by the action of strong alkalis Darkens with age Limited water resistance	Mixed with red or white lead for metal and timber primers Oil-modified resin mediums Chemically combined with alkyd, epoxy and urethane resins
BOILED OIL (linseed oil heated with driers added)	Dark brown in colour Dries within 12 hours Improved water-resistance, compared with linseed oil Good adhesion Saponified by alkalis Darkens and becomes brittle with age	Oil-modified resin mediums
STAND OIL (partially polymerised oil)	Pale yellow in colour High viscosity Excellent flow Good flexibility Not alkali-resistant	Oil-modified resin mediums Some wood and metal primers Manufacture of alkyd resins
SOYA BEAN OIL	Pale yellow Very slow-drying Does not darken with age Not alkali- or water-resistant	Combined with resins, particularly in pale and non-yellowing paints
TUNG OR CHINA WOOD OIL	Milky white in colour Drying time similar to linseed oil Resistant to alkali and water	Combined with resins for use in alkali-resisting and water-resisting coatings
TOBACCO SEED OIL	Slow-drying	Chemically combined with alkyd resins
OITICICA OIL	Similar properties to tung oil	Sometimes used as an alternative to tung oil
DEHYDRATED CASTOR OIL	Can dry in under 5 hours Pale colour, non-yellowing	Chemically combined with alkyd and epoxy resins

TABLE 2.4   **Properties of natural resins**

Name	Properties	Use
COPAL Fossil resins obtained mainly from the Congo	Very hard Pale colour	Oil varnishes
MANILA COPAL Fossil resins obtained from Manila	Bleed-resistant	Road-marking paints
KAURI Fossil resin obtained from New Zealand	Pale colour Combines well with oil	Oil varnishes
DAMAR Recent resin tapped from living trees in Malaysia	Very flexible	Cellulose lacquers Shellac varnishes
LAC Recent resin obtained from excreta of tree-feeding insects in India	Soluble in alcohol Bleed-resistant	Knotting Button polish French polish

TABLE 2.5   **Properties of synthetic resins**

Name	Properties	Use
OIL-MODIFIED ALKYD RESINS Containing over 60% of a drying oil	Good flow High gloss Good flexibility Weather-resistant Pale colour Not alkali-resistant	Some air-drying primers Most undercoat, gloss and eggshell finishes Varnishes
Modified to a gel or thixotropic structure	Good brushability High gel strength Fast gel recovery rate Heavy film build Non-drip	Thixotropic paints One coat paints
Containing non-drying oils	Very slow/non-drying Pale colour Soft	Plasticiser in cellulose paints
EPOXY OR EPOXIDE RESINS 2-pack or 'cold cure' type	Excellent adhesion Excellent water-resistance Excellent chemical-resistance Excellent abrasion-resistance Extremely hard film	Chemical-resistant coatings Abrasion-resistant coatings Water-resistant coatings
1-pack epoxy esters (modified with drying oils)	Chemical-resistant, but much less than 2-pack types	Maintenance paints for factories

TABLE 2.5   **Properties of synthetic resins** – *continued*

Name	Properties	Use
POLURETHANE RESINS 2-pack or 'cold cure' type	Extremely hard film Excellent abrasion-resistance Excellent weather-resistance Excellent chemical-resistance but less alkali-resistant than epoxy	Chemical-resistant finishes Abrasion-resistant coatings
1-pack based on drying oils (urethane oil)	Films are harder, tougher and more water-resistant than most alkyd resins	Interior and exterior gloss finishes Clear wood finishes
POLYVINYL ACETATE (PVA) Copolymer	Good colour, non-yellowing Good flowing properties Good adhesion Good water-resistance Alkali-resistant Good external durability Good washability	Adhesives Emulsion paints Masonry paints
ACRYLIC EMULSIONS	Excellent adhesion Water-white colour Non-yellowing Very good washability Alkali-resistant Very good external durability	Emulsion paints Timber primers Quick-drying undercoats Adhesives Masonry paints
PHENOLIC RESINS	Very good water-resistance Alkali-resistant Poor colour Yellows badly	Alkali-resistant varnishes Marine varnishes Anti-corrosive primers
COUMERONE RESIN	Alkali-resistant Low acid value Tends to yellow	Alkali-resistant paints Metallic paints
MALEIC RESIN	Very pale in colour Non-yellowing Good gloss Poor resistance to chemicals	Combined with alkyd resins for non-yellowing paints Pale varnishes
UREA/FORMALDEHYDE RESIN (2-pack acid catalysed)	Very pale colour Non-yellowing Hard glossy films Heat-resistant	Clear lacquers for bar counters and furniture
SILICONE RESIN	Water-resistant Heat-resistant up to 475°C	Clear waterproofing solutions Heat-resisting paints
POLYVINYL BUTYRYL	Excellent adhesion	Etch primers

2.1.3 FILM-FORMERS

The film-formers used in modern surface coatings can be broadly divided into four groups: drying oils, resins, oil-modified resins and other film-forming substances.

Drying oil (Table 2.3)
Drying oils are vegetable oils which, when exposed to air, convert from liquid to solid. The conversion (they are known as convertible coatings) is a slow process: the oil absorbs oxygen, and slowly thickens until it dries to a solid film. This drying process is called oxidation. Drying oils are occasionally used as the sole film-former but more often are mixed with resins (see resins and oil-modified resins).

Resins
Resins form the basis of most film-formers and provide the properties required by the range of coatings described in Sections 2.2–2.5. They are obtained from two main sources.
Natural resins (Table 2.4) are obtained from trees and may be either (a) extracted from living trees or (b) fossil resins from the remains of trees, mostly tropical, long extinct and lying fossilised in swampy areas in many parts of the world. Because the supply is limited, the cost of collecting them is rising and the quality varies considerably. Natural resins have a limited use in the paint industry.
Synthetic resins (Table 2.5) are manufactured under strict quality control to specifications designed for a precise purpose. They may be either (a) organic, obtained from carbonaceous materials, the chief source being crude petroleum oils; or (b) inorganic: only one resin

commonly used in paint manufacture, silicone resin, is obtained from material such as sand and quartz.

Most surface coatings contain a synthetic resin-based film-former. Many decorative paints are based on oil-modified resins; these are not mixtures of oils and resins, but the result of a chemical combination of the two during the manufacture of the resin.

Oil-modified resins (Table 2.6)
Combinations of oil and resin are usually carried out during the manufacture of the resin, the resultant oil-modified resin being chemically and physically stable. A few mixtures of oil and resin are still used. These are prepared by melting the resin and stirring in the oil.

Other film-forming substances (Table 2.7)
A few organic substances other than oils and resins are used in some paints and varnishes.

2.1.4 THINNERS AND DRIERS

Thinner (solvent/diluent)
Thinners are colourless and are selected according to the type of paint in which they are employed. They are incorporated into the liquid part of the paint but evaporate from the film during the drying process.
Properties
 (i) Act only on the medium of the paint.
 (ii) Remain long enough in the wet paint to allow easy application.
 (iii) Are free from residue which may affect the film.
 (iv) Are colourless and do not cause discolouration.

TABLE 2.6 Properties of oil-modified resins

OIL-MODIFIED (NATURAL) RESINS (natural resin and linseed or other drying oil)	High gloss Flexible film Saponified by alkalis Limited water-resistance Yellows on ageing	Some primers Some undercoats Varnish Gold size
OIL-MODIFIED (SYNTHETIC) RESINS (synthetic resin and tung or other oils)	Water-resistant Alkali-resistant	Acid- and alkali-resistant paints and varnishes (limited resistance) Boat varnish

TABLE 2.7   **Properties of other film-forming substances**

Medium	Properties	Use
BITUMENS (asphalt, bitumen and pitch)	Very dark colour Excellent water-resistance Resistant to acids Resistant to alkalis Low cost Colour bleeds through oil/resin paints Softens on heating	Damp-proof compounds Bituminous paints
RUBBER (natural and synthetic, treated to form chlorinated rubber, isomerised rubber, cyclised rubber, neoprene, hypolon)	Excellent water-resistance Chemical-resistant Very flexible Quick-drying Good flow	Chemical-resistant paints Water-resistant paints
CELLULOSE (nitro-cellulose: cellulose treated with nitric acid)	Rapid drying Hard film Chemical-resistant Almost impossible to brush Tends to be brittle	Cellulose paints and lacquers Cellulose stoppers and fillers

(v) Are not strong enough to soften underlying coatings.

Use   When used in decorative paints their main use is to reduce the viscosity of primers, fillers and undercoats, making them easier to apply and ensuring penetration into absorbent surfaces. Thinning should only be carried out to manufacturers' instructions, as over-thinning can lead to poor opacity, loss of gloss and other film weaknesses.

Types of solvent   Seven main classes of solvent are used in paint:

Hydrocarbons	Ketones
Alcohols	Chlorinated hydrocarbons
Esters	Water
Ethers	

Hydrocarbons

(a) Those obtained during the distillation of crude oil: petrol, white spirit and paraffin.

(b) Aromatic hydrocarbons, obtained from the distillation of coal tar: solvent naphtha, benzine, toluene and xylol. These are used in chlorinated rubber and nitro-cellulose paints.

Alcohols

These used to be produced by distilling fermented potatoes and starch. Industrial alcohols are now produced by a chemical process.

Propyl alcohol, butyl alcohol, amyl alcohol, benzyl alcohol, isopropanol alcohol, ethyl alcohol and methylated spirits are commonly used for dissolving resins and as thinners for spirit paints and varnishes.

Esters

These are derivatives of acids and include methyl acetate, ethyl acetate, butyl acetate and amyl acetate. They are used as solvents for nitro-cellulose finishes.

Ethers

Prepared by chemical action on certain other solvents. Glycol ether and methyl glycol are used as solvents for cellulose resins and many other synthetic finishes.

Ketones

Manufacture is similar to that of alcohols. They include acetone, methyl ketone and methyl ethyl ketone, and are used extensively in nitro-cellulose finishes.

Chlorinated hydrocarbons

Produced by a reaction of chlorine on various materials such as methane. Their main disadvantage is toxicity, but they have the advantage of being non-flammable. Two main types are in use: (a) methylene chloride, a very strong solvent used as the main constituent in paint removers; (b) trichlorethylene, used as a degreasing agent.

Both types give off poisonous gases when burnt.

Water

All emulsion paints are thinned with water. Water is cheap, readily available, non-flammable and non-toxic.

Flash point

Flash point is the lowest temperature at which a substance will give off flammable vapour. With the exception of chlorinated hydrocarbons and water, all the above solvents are flammable, having different flash points. (See Section 6.6.)

Drier

Driers are additives to paint, used to speed up the drying process. They increase the capacity of the film-former to absorb oxygen from the atmosphere. An oil paint which takes approximately four days to dry without driers will dry overnight with the addition of one per cent drier.

Properties

(i) Speeds up the drying process.

(ii) Mixes easily with drying oil.

Use Normally incorporated by the manufacturer into the drying oil. Can be used by the decorator in oil colours for signwriting or scumble and glazes where driers are not normally added.

Where driers are used they should only be added very sparingly as over-use will lead to cracking or shrivelling.

Composition Driers are soluble compounds of metals, most commonly manganese, cobalt and lead, mixed with other ingredients. The two main types are naphthenates and octanoates.

The metals have their own characteristics and are often used in combination to give the desired properties.

Types The decorator can obtain driers in the form of (a) liquid drier, consisting of a solution of lead and cobalt naphthenates in white spirit; (b) terebine drier, a mixture of lead and manganese with linseed oil, rosin and white spirit which produces a brittle film.

2.1.5 PROPERTIES OF PAINT

By the careful selection and mixing of the film-formers, pigments and thinners available, the paint manufacturer can produce materials that under most known conditions will satisfy the requirements of four reasons for painting:

(a) to protect the surface (*preservation*)

(b) to make the surface washable (*sanitation*)

(c) to decorate the surface (*decoration*)

(d) to identify the surface (*identification*).

In addition to these properties, paint must have certain basic qualities to ensure that it (i) is appliable; (ii) will dry in a reasonable length of time; (iii) can adapt to the physical changes of the surface; (iv) will maintain its function for an acceptable period.

Consistency

This can also be referred to as the fluidity of paint. Certain trade terms are used to describe it, such as 'round' for thick paint, 'sharp' for thin paint and 'viscous' for syrupy or sticky paint.

Consistency depends on:

(a) *Type of film-former:* gloss paints and varnishes feel 'sticky' when being applied, compared with the smooth feel of emulsion paint. The consistency of paint changes as the thinner evaporates or the set begins to take place.

(b) *Type of pigment:* coarse pigments drag on the brush, therefore influence the paint's fluidity; for example, a masonry paint which contains granules of stone is much more difficult to brush than one containing no coarse aggregate.

(c) *Temperature:* an ideal temperature for painting is 15°C. Below this temperature the resins and oils become more viscous and 'drag' on the brush or roller. Paints become very difficult to apply at temperatures below 5°C. Higher temperatures can also affect brushing and rolling consistency as the thinner evaporates quickly, making the paint less fluid.

(d) *Thixotropy:* many paints are deliberately made in a very 'round' condition and are called thixotropic or *gel* coatings. These include a component that gives the paint the qualities of thickness in the container but fluidity when brushed or rolled. Thixotropic paints allow thicker films to be applied.

(e) *Surface condition:* absorbent surfaces readily absorb the liquid part of the paint, so that brushing and rolling may become difficult.

Opacity (hiding power)

Opacity is the property of a paint to obliterate the surface to which it is applied. Although the pigment is the principal component responsible for opacity, a careful blending of appropriate film-formers and pigments is also important. Many strong-coloured paints have poor opacity, mainly because they contain no white base pigments and the coloured pigments that produce the hue have poor opacity. Many yellow, orange and red paints are almost transparent.

Spreading capacity

This is defined as the area covered by a given quantity of paint. It is influenced by both the film-former and the pigment and is closely related to paint consistency; for example, a smaller area will be covered by a litre of heavy-bodied or 'sticky' paint than of a thin or smooth paint. Rough, textured or absorbent areas will require a greater quantity of paint to cover them than will a smooth surface.

Adhesion

The resin in the paint provides its ability to stick to a surface. An alkyd gloss paint will adhere to a smooth plastic surface better than a spirit varnish.

The condition of the surface is vitally important to the maximum adhesion of the paint film.

Elasticity (flexibility)

Timber and metal surfaces naturally expand and contract throughout their life. The degree of movement depends upon temperature and humidity, and varies considerably. Paints applied to such surfaces must be able to stretch and shrink to the same degree. Most oil and oil/resin film-formers have this ability. Some surface coatings, such as shellac knotting, are brittle and readily crack when applied to flexible substrates.

Density

Pigments with high specific gravity tend to settle out of paint to form a solid mass at the bottom of the container. They also affect the spreading property of the paint as they tend to 'drag'. The most obvious effect of dense pigment is on the weight of the paint: a litre of lead primer will be considerably heavier than a litre of leadless primer.

Drying speed

Paints are manufactured to 'stay open' long enough for easy and even application. The hardening process is much longer and varies according to the type of film-former used. Oil paints set-up in less than $\frac{1}{2}$ hour, and touch-dry in 4–6 hours, can be overcoated after 12 hours but do not attain maximum hardness for many weeks. Emulsion paints have a similar set-up time, but are touch-dry after 1–2 hours and can be recoated in 2–4 hours.

Another important aspect of drying speed is a paint's 'wet-edge' time. This is similar to its set-up time, and is the length of time that an applied coat takes to become so tacky that the next section cannot be brushed into it without showing a join. It is a characteristic of paint which must be accurately judged when covering large areas.

Flow

Flow is the quality of paint that enables the painter to apply coatings so that brush or roller marks are not visible. Some paints, such as gloss paints, have good flowing properties and help the painter to produce a texture-free finish. Other paints, such as undercoats and flat finishes, have little flow and require different application techniques to produce an acceptable finish.

Durability

Correctly specified paint systems, applied skilfully to sound, properly prepared surfaces will provide the required protective, washable or decorative coating. The effective life of the coating depends principally on (a) the destructive effect of the environment, i.e. air pollution, dampness, salt spray, ozone, ultraviolet light, industrial chemicals, heat and cold; (b) the natural breakdown of the film-former as it ages and becomes more and more brittle. A paint which may last for ten years in a rural area may break down in two years in a heavily polluted industrial area. An average life may be stated as five years.

2.2 PRIMERS, UNDERCOATS AND STOPPERS

2.2.1 ACRYLIC PRIMER/ UNDERCOAT

A water-thinned paint made to fulfil the functions of a combined primer and undercoat. Available mainly in white, off-white and pink; some manufacturers supply other pastel shades.

Composition

Pigment Titanium white, extenders and coloured pigments.

Film-former Mainly acrylic resin.

Thinner Water.

Properties

Quick drying enables same day overcoating.

Water-based: cleaning and thinning can be carried out quickly and cheaply.

Hard, flexible film.

Good adhesion to prepared surfaces.

Alkali-resistant.

Non-toxic.

Non-reversible coating.

Use Can be applied direct to the following interior or exterior surfaces: new and old plaster, cement, concrete, asbestos, woodwork, hardboard, building boards, brickwork. Not to be used over bare metal. Steel hinges, nail heads and other bare steel fixings should be 'spot primed' with a metal primer and allowed to dry before applying acrylic primer. Can be used as a matt finish on interior ceilings and walls.

Application

Drying

Method By coalescence.

Time 30 minutes normally, but in cold damp conditions the film may remain wet for a number of hours.

Overcoating As soon as the film is touch-dry.

Cleaning solvent Clean water.

Spreading capacity 65–90 m^2 per 5 litre, depending upon the porosity of the surface.

Note Has poor flowing properties.

2.2.2 ALKALI-RESISTING PRIMER

Primer specially designed for surfaces which are alkaline in nature.

Composition Based on resins and pigments which are resistant to alkali, such as:

Pigment Titanium and china clay.

Film-former Coumarone and phenolic resin and tung oil.

Thinner White spirit.

Properties

Alkali-resistant.

Resistant to limited water pressure from inside the substrate. Non-reversible coating.

Use

 (i) Primer on all new and old building materials of an alkaline nature (see 3.4.5).

 (ii) Primer for surfaces which have been coated or impregnated with fire-retardant solutions.

(iii) Binding down powdery surfaces.

Application

Brush recommended for maximum adhesion. On smooth surfaces one full coat is sufficient but on very porous or textured surfaces two coats may be required.

Drying

Method Oxidation.

Time Under normal conditions, 3–4 hours.

Overcoating Should be left overnight before overcoating. Can be overcoated with any decorative system but if water-based materials are used some cissing may occur due to the oily nature of the primer. It is advisable not to rub down the surface before overcoating, to avoid breaking the surface.

Cleaning solvent White spirit.

Spreading capacity Depends entirely on the porosity and texture of the surface but the following will give a general guide.

	m^2 per 5 litres
Smooth hard plaster surfaces	45–55
Cement rendering slightly textured	35–45
Stucco, rough-cast	15–30

2.2.3 ALUMINIUM WOOD-PRIMER

A dull metallic grey oil-based primer.

Composition

Pigments Non-leafing aluminium paste. Non-leafing aluminium particles are scattered throughout the film at various angles and provide good build and adhesion for subsequent coats (Fig. 2.4).

Fig . 2.4 *Structure of aluminium wood-primer*

Film-former Tung oil/phenolic resin. Tung oil/coumarone resin.

Thinner White spirit.

Properties

Good sealing properties over surfaces which are likely to bleed.

Good opacity.

Very good spreading capacity.

Self-knotting.

Good moisture-barrier.

Heat-reflecting.

Lead-free.

Non-reversible coating.

Use

(i) Wood-primer specially suitable for resinous timber such as Columbian pine, Oregon pine, afrormosia, gurgion and iroko.

(ii) Sealer over surfaces previously treated with coal-tar wood preservatives.

(iii) To seal old bitumen-coated surfaces, but where thick layers of bitumen are present cracking may occur.

Application

Drying

Method Oxidation.

Time 4–6 hours. Because of the laminar structure of the pigment the lower parts of the film may remain soft after the surface has dried, due to trapped solvents and lack of oxygen.

Overcoating Allow 24 hours before applying a decorative system.

Cleaning solvent White spirit.

Spreading capacity 80 m^2 per 5 litre.

Note Because of its colour an extra undercoat may be required if a light coloured finishing coat is used.

2.2.4 CALCIUM PLUMBATE PRIMER

Good all-purpose primer, off-white or grey in colour. Designed for use as a first coat on galvanised iron.

Composition

Pigment Should contain not less than 33 per cent lead in the form of calcium plumbate mixed with small quantities of titanium, asbestine or barytes.

Film-former   Linseed oil.

Thinner   White spirit.

Properties

Can be used on exterior or interior work.

Excellent rust-inhibition.

Prevents the spread of corrosion on ferrous metals where paintwork is scratched or broken.

Excellent adhesion.

Good opacity.

Good flow and levelling properties.

Non-reversible coating.

Use

(i) Specially suited for priming galvanised iron without the need for weathering.

(ii) Primer for iron and steel where light colours are used.

(iii) Priming hardwoods such as oak. Composite units, e.g. wood and metal window frames, can be primed with the same primer.

Application

Drying

Method   Oxidation.

Time   Touch-dry in 6–8 hours.

Overcoating   The primed surface should be left for 24 hours before overpainting.

Cleaning solvent   White spirit.

Spreading capacity

Woodwork   40–60 m^2 per 5 litre.

Metalwork   50–60 m^2 per 5 litre.

Note   Not recommended over aluminium; not suitable for immersion in water for long periods; toxic.

2.2.5 ETCH PRIMER

Designed for pretreatment of clean ferrous and non-ferrous metals to ensure maximum adhesion of the paint system. Available in one-or two-pack materials, the latter offering greater stability and adhesion. Sometimes called wash primers.

Composition   Both types have the same basic components in differing proportions. The main difference is that in the two-pack paint the acid is supplied separately and added just before use. The main components are:

Pigment   Zinc tetroxychromate.

Film-former   Polyvinyl butyryl resin.

Thinner   Isopropyl alcohol.

Activator   Phosphoric acid.

Plasticisers and extenders are also added.

Properties

Etch primers work on a principle of interaction between pigment, binder and substrate, resulting in excellent adhesion.

Can be used on composite units where different types of metal are used.

Matt finish, giving good key.

Corrosion-resistant.

Lead-free.

Non-reversible coating.

Use   For pretreatment of untreated aluminium, galvanised iron, zinc, metal-sprayed steel, copper, brass, bronze, lead, tin, cadmium and chromium plating, clean iron and steel.

Note   Should not be used where a phosphoric acid wash has been used.

Application

They should be applied in dry conditions as they are sensitive to water.

Drying

Method   Polymerisation.

Time   Touch-dry in 1–4 hours.

Overcoating   A metal primer should be used over etch primers after the recommended hardening time. Water should not be allowed to contact the surface before overcoating is carried out.

Cleaning solvent   Only the manufacturer's recommended cleaning solvent should be used.

Spreading capacity   80–90 m^2 per 5 litre. On abrasive-cleaned or cast metals spreading capacity will be reduced owing to the greater surface area.

Note   Limited shelf life; once mixed, two-pack etch primers have a pot life of 8–24 hours; they give no protection if left unpainted; provide a very thin semi-transparent film.

2.2.6 LEAD-BASED WOOD-PRIMER

An excellent primer for exterior woodwork, available in white or pink. Although still available for special specifications requiring lead paint, it is almost obsolete due to the toxic lead content. Being replaced with lead-free primers.

Composition
Pigment White lead and titanium white with a small quantity of red lead if a pink primer is required.
Film-former Linseed oil or other resin-modified oil varnish.
Thinner White spirit.
Properties
Good adhesion.
Good penetration.
Flexibility.
Impervious to water.
Non-reversible coating.
Toxic.
Use
  (i) Restricted to outside use for toxic reasons.
 (ii) Used particularly for woodwork which is exposed to severe weather conditions, sea spray, etc.
(iii) Should be used for back-priming on joinery.

Application
Drying Method Oxidation.
Time 6–10 hours.
Overcoating After overnight drying.
Cleaning solvent White spirit.
Spreading capacity 60–70 m² per 5 litre.
Note Persons working with lead paints must observe the Control of Lead at Work Regulations (see 6.4.1); these primers are being superseded by lead-free primers for health reasons.

2.2.7 LEAD-FREE WOOD-PRIMER

A general-purpose wood-primer available in white and pink.

Composition
Pigment Titanium and lithopone, paris white and barytes.
Film-former Oil-modified resin varnish. The composition of the film-former can be adjusted to suit the requirements of the film; i.e. for outside use, a long oil resin would give a more flexible film than a short oil resin, which would dry quicker and harder, fulfilling the requirements of an interior primer.
Thinner White spirit.
Properties
Satisfies the porosity of the wood.
Provides good adhesion for succeeding coats.
Non-toxic.
Not as water-resistant as white lead.
Non-reversible coating.
Use
  (i) Interior work where children or pets may chew the surface.
 (ii) Areas where sulphurous fumes are present which would discolour white lead paints.
(iii) Nurseries, hospitals and places where food is stored or manufactured.
Application

Brushing ensures better penetration.
Drying
Method Oxidation.
Time Touch-dry in 4–8 hours.
Overcoating Any decorative paint system can be used. Overnight drying should be allowed.
Cleaning solvent White spirit.
Spreading capacity 50–60 m² per 5 litre.

2.2.8 SPIRIT-BASED ALUMINIUM SEALER

Designed for use over surfaces which have a tendency to soften or bleed through subsequent coats of paint.
Composition The ingredients of these paints should be of a type that will not disturb or dissolve the surface or part of the surface to which they are applied.
Pigment Aluminium flake non-leafing.

Film-former Shellac or manila gum.
Thinner Industrial methylated spirit.
Properties
Forms barrier film to insulate bleeding surfaces.
Quick-drying.
Good opacity.
Resistant to mild chemical fumes.
Seals water-stained surfaces.
Reversible coating.
Use
 (i) Sealer for bituminous surfaces, dry creo-soted surfaces and nicotine stains.
(ii) Sealer for bleeding colours or inks in previously painted or papered surfaces.
Application

Drying
Method Evaporation.
Time 15–30 minutes.
Overcoating Allow 1 hour before applying finishing system or papering.
Cleaning solvent Methylated spirit.
Spreading capacity Approximately 30–35 m^2 per 5 litre.

2.2.9 PENETRATING OILY SEALER OR STABILISING SOLUTION

Designed for use on powdery or unstable surfaces to penetrate and bind loose particles together. It may be either clear or pigmented, the clear having better penetrating properties.
Composition
Pigment Sometimes added to give opacity.
Film-former Oil-modified resin based on tung oil, phenolic or coumarone resin. The varnish: solvent ratio is carefully controlled to produce a good balance. If the sealer is too thick it will lie on top of the surface, leaving loose material underneath; if too thin, it will penetrate too far into the surface, leaving unbound material on the surface resulting in poor adhesion.
Thinner White spirit.
Properties
Binds powdery surfaces.
Seals porous surfaces.

Will take any form of decoration.
For inside and outside use.
Clear or pigmented.
Non-reversible coating.
Use
 (i) To stabilise old powdery, highly porous, limewashed, distempered or cement-painted ceilings and walls. These sealers are not intended to take the place of adequate preparation (see 3.3).
(ii) To seal plasterboard before paperhanging, to avoid damage to the board when stripping.
(iii) Clear sealer may be used to seal loose screeds and powdery floors.
Application

Drying
Method Oxidation.
Time Approx. 4 hours.

2.2.10 RED LEAD PRIMER

A rust-inhibitive primer designed to protect iron and steel from corrosion. Bright red in colour. Although still available for special specifications requiring lead paint, it is almost obsolete due to the toxic lead content. Being replaced with lead-free primer such as zinc phosphate metal primer.
Composition Red lead pigment in a linseed oil medium, thinned with white spirit.
Properties
Excellent rust-inhibition.
High film-build.
Can be used effectively over hand-cleaned surfaces without significant loss of performance.
Non-reversible coating.
Use
 (i) Heavy structural steelwork such as industrial plants, installations, gas-holders and storage tanks.
(ii) Light structural steel such as railings, rainwater pipes and gutters.
Application

Drying
Method Oxidation.
Time 2–4 hours, may take longer in damp or cold conditions.
Overcoating The primer should be left to dry for at least 24 hours before overcoating. Excellent results are obtained when overcoated with micaceous iron oxide paint. Most oil-based decorative systems can be used over red lead primer.
Cleaning solvent White spirit.
Spreading capacity 60–85 m² per 5 litre.
Note Has poor flow and levelling properties; toxic; use controlled by Control of Lead at Work Regulations (see 6.4.1).

2.2.11 ZINC CHROMATE METAL-PRIMER

A non-toxic yellow metal-primer, sometimes referred to as a universal metal-primer.
Composition
Pigment Zinc chromate (red oxide is sometimes added).
Film-former Alkyd or oil-modified resin.
Thinner White spirit.
Properties
Rust-inhibitive.
Good flow and levelling properties.
Good resistance to strong solvents which may be present in subsequent coats.
Resistant to marine atmospheres.
Can be sprayed.
Non-toxic.
Suitable for interior or exterior.
Non-reversible coating.
Use
 (i) Clean or etched aluminium and light alloys, sprayed zinc of sherardised surfaces.
 (ii) Ferrous metals.
(iii) Composite wood and metal components.
Application

Drying
Method Oxidation.
Time 16–24 hours.
Overcoating Should be carried out as soon as possible after the primer is dry: approximately

24 hours. Most types of decorative and industrial paints can be used, including chlorinated rubber.
Cleaning solvent White spirit or manufacturer's recommended cleaning solvent.
Spreading capacity 55 m² per 5 litre.

2.2.12 ZINC PHOSPHATE METAL-PRIMER

A special rust-inhibitive primer developed as an alternative to red lead.
Composition
Pigment Zinc phosphate and red oxide.
Film-former Resin-modified drying oil.
Thinner White spirit.
Properties
Excellent rust inhibition.
Non-toxic.
High film build (dry film 60–80 microns).
Non-reversible coating.
Use Suitable for all ferrous metal surfaces under oil-based paint finishing systems. (Not suitable for immersion in water.)
Application

Drying
Method Oxidation.
Time Recoatable in 16 hours – longer in cold, damp conditions.
Cleaning solvent White spirit.
Spreading capacity 40–50 m² per 5 litre.

2.2.13 ZINC-RICH PRIMER

A special primer based on pure zinc dust, sometimes referred to as 'cold galvanising'.
Composition 90–95 per cent (by weight) pure zinc dust dispersed in a medium such as chlorinated rubber or epoxy resin.
Properties
The film is 'sacrificial', i.e. the zinc corrodes instead of the steel (cathodic protection).
The zinc particles, when in contact with each other and the bare steel, will gradually cold-weld themselves together.
Excellent water-resistance.

Will protect scratched or broken films against further corrosion.

Not essential to overcoat immediately as it can protect a surface for many years.

Film thickness 50–75 microns per coat (average thickness of one coat of paint is 25 microns).

Non-reversible coating.

Use

(i) Ideal where speed of drying is important, e.g. between tides on marine installations.

(ii) For factory-made units which may be exposed during transport and storage on site.

Application Preparation of the metal should be by blast cleaning. Must not be applied over old paint and not recommended for 'spot priming'. Small areas can be brush-painted.

Drying

Method Chlorinated rubber, by evaporation; epoxy resin, by polymerisation.

Time Epoxy-based types can be handled after only 5–10 minutes; other types may take 30–60 minutes.

Overcoating The primer can be overcoated after a few hours but it is advisable to leave overnight before recoating. A wide variety of paint systems may be applied over the primer but it is always advisable to consult the manufacturer for guidance.

Cleaning solvent The manufacturer's recommended cleaning solvent, immediately after use.

Note Readily attacked by acids; very expensive.

2.2.14 OIL/RESIN UNDERCOAT

A heavily pigmented paint which dries to a semi-matt finish. Designed for use under gloss paints.

Composition

Pigment Titanium white with extenders and coloured pigments as required.

Film-former Oil-modified alkyd or drying oil.

Thinner White spirit.

Properties

Good adhesion to the primer.

Good opacity.

Good flow and levelling.

High film-build.

Of lighter tone than the finish.

Non-reversible.

Use To give body and colour to a paint system. Can be used over all suitably primed surfaces, inside or outside.

Application

Drying

Method Oxidation.

Time 6–10 hours.

Overcoating Overnight or 12–16 hours. Paints containing strong solvents and some eggshell paints may disturb and lift the dried film and should not be used over oil undercoats.

Cleaning solvent White spirit.

Spreading capacity 80–90 m^2 per 5 litre.

2.3 FINISHING PAINTS

2.3.1 BRONZE/ALUMINIUM PAINT

Available in a wide range of metallic colours, ranging from bright aluminium to dull bronze.
Composition Bronze or gold paints are made from mixtures of zinc and copper in varying proportions to give a number of different colours. Silver paint is made from aluminium dust in leafing form. Both types can be dispersed in the following mediums:
(a) Cellulose lacquer
(b) Oil/varnish
(c) PVA emulsion glaze.
Properties
Sealing properties over surfaces likely to bleed.
Light-reflectance.
Bright metallic lustre.
Heat-resistance.
Use
Aluminium Lamp-posts, roof trusses, structural steel and hot water pipes.
Bronze Decorative finish for shop-fronts, wrought ironwork and for relief decoration.
Application

Drying
Method Cellulose type, by evaporation; PVA type, by coalescence; oil type, by oxidation.
Time Because of the leafing properties of the pigment, solvents may take some time to evaporate. Drying of oil types can take up to 12 hours.
Overcoating Overnight drying should be allowed before further coats are applied.
Cleaning solvent Cellulose thinner, white spirit or water. Brushes should be well cleaned in solvent, then detergent, and finally rinsed in clean water to ensure that all particles of metal are removed.
Spreading capacity 90–100 m² per 5 litre.

2.3.2 EGGSHELL OR SEMI-GLOSS FINISH

An interior decorative finish which dries with a sheen. Also known as satin, silk or lustre finish.
Composition
Pigment Titanium white and a wide range of coloured pigments.
Film-former Oil-modified alkyd resin (some types reinforced with polyurethane for extra hardness)
Properties
Thixotropic.
Good flow.
Washability.
Non-toxic.
Easily cleaned with detergent.
Film tougher and more washable than emulsion paints.
Requires no undercoat.
Available in wide range of colours.
Excellent opacity (except some yellows and oranges).
Resists normal domestic steam and heat.
Non-reversible.
Use Decorative finish for interior surfaces including ceilings, walls, woodwork and metal frames.
Application

Drying
Method Oxidation.
Time Touch-dry in 4–6 hours.
Overcoating Overnight or 16 hours.
Cleaning solvent White spirit (with some types brushes can be cleaned in hot detergent solution).
Spreading capacity 60–80 m² per 5 litre.

2.3.3 MATT EMULSION PAINT

A water-thinned material suitable for ceilings and walls.
Composition
Pigment Titanium white, extenders and a wide range of coloured pigments.
Film-former Polyvinyl acetate (PVA) copolymers or acrylic polymers or a combination of the two.

Properties
Permeable: moisture vapour will pass through the film.
Excellent adhesion.
Flexible films.
Washable.
Available in a wide colour range, or clear.
Alkali-resistant.
Good sealing properties over bituminous materials.
Non-toxic.
Application quicker and easier than oil paints.
Non-reversible coating.
Use Extensively used for walls and ceilings. Suitable for use over plasterboard, fibre insulating board, wood, hardboard, plaster, brick, cement rendering, stucco, asbestos sheeting and fabrications, foamed polystyrene and wallpaper.
Application

Drying
Method Coalescence.
Time 1–2 hours.
Overcoating Allow 4 hours. Emulsion painted surface can be papered or painted over after cleaning and denibbing.
Cleaning solvent Water. Cleaning should be carried out immediately after use.
Spreading capacity 60–65 m^2 per 5 litre on smooth non-porous surfaces.
Note The dry film can be softened by methylated spirit.

2.3.4 VINYL SILK EMULSION

Similar to emulsion paint but with poorer opacity and drying to an eggshell sheen.
Composition
Pigment Titanium white with wide range of coloured pigments.
Film-former Copolymer vinyl or acrylic polymer.
Thinner Water.
Properties Semi-gloss finish.
Alkali-resistant.
Non-toxic.
Washable.

Non-tainting.
Non-yellowing.
Non-reversible coating.
Use Where the properties of emulsion paints are required but with a higher sheen, e.g. in kitchens, bathrooms, hospitals, schools, food preparation plants.
Application

Drying
Method Coalescence.
Time 1–2 hours.
Overcoating 2–4 hours.
Cleaning solvent Warm water.
Spreading capacity 60–65 m^2 per 5 litre.

2.3.5 GLOSS FINISH

Interior and exterior decorative paint having a full gloss finish. Used as the main protective coating in the decorating industry.
Composition
Pigment Titanium white and a wide range of light-fast pigments to produce the full colour range.
Film-former Alkyd resins modified with drying oils. Some types contain a small proportion of other resins such as silicone and polyurethane resins to improve the general properties.
Thinner White spirit.
Properties
Excellent gloss.
Excellent flow.
Good flexibility.
Very good weather-resistance.
Heat-resistant up to 93°C (suitable for hot water radiators).
Non-reversible coating.
Use General interior and exterior finishes. Decorative finish over suitably primed and undercoated timber, metals, plaster, concrete, brickwork and building boards.
Application

Drying
Method   Oxidation.
Time   Tack-free in 4–6 hours.
Overcoating   Leave overnight to dry.
Cleaning solvent   White spirit.
Spreading capacity   75–85 m² per 5 litre.

2.3.6 MULTI-COLOUR FINISH

An interior decorative finish which has 2, 3 or 4 separate colours in the form of spots, flecks or streaks applied in one application. Each fleck of paint exists as an individual both in the tin and during application. Then as the film dries it joins up with its neighbours to form a multi-colour finish. The paints contain two incompatible materials, which allows several colours to be used in one container without becoming mixed together. Three types are available:
(a) Cellulose/water medium (spray applied).
(b) Vinyl/resin medium (spray applied).
(c) A brush-applied clear coating containing coloured flecks applied over an opaque coloured ground.

(a) Cellulose multi-colour finish
Composition
Pigment   Titanium white and range of coloured pigments.
Film-former   Pigments ground in cellulose medium dispersed in stabilised water.
Properties
Anti-static, does not attract dust.
Non-toxic.
Washable.
Does not chip.
Excellent adhesion.
Camouflages surface irregularities and joints in boards.
Non-reversible coatings.
Hardwearing and abrasion-resistant.
Liable to lift oil paint substrates.
Shelf-life 6–9 months.
Strong odour and fumes which can be unacceptable in hospitals, food factories, occupied buildings.
Use   All types of interior decoration where a decorative and hardwearing surface is important, e.g. in toilets; corridors in schools, hospitals and public buildings. Suitable for application over glazed tiles.

Application

Medium or large set-ups give best results. Pressure should be about 20–30 psi (1.5–2 bars).
Drying
Method   Evaporation.
Time   Touch-dry in 2 hours; hard overnight.
Cleaning solvent   Water, and the manufacturer's cleaning solvent.
Spreading capacity   16–20 m² per 5 litre.

(b) Vinyl multi-colour finish
Composition
Pigment   Titanium white and range of coloured pigments.
Film-former   Pigments ground in low-odour spirit-carried vinyl resin dispersed in stabilised water.
Properties
Anti-static.
Non-toxic.
Washable.
Does not chip.
Excellent adhesion.
Camouflages surface irregularities and joints in boards.
Non-reversible coatings.
Hardwearing and abrasion-resistant (although inferior to cellulose multi-colour).
Can be applied over oil paint substrates.
Shelf-life of 6–9 months.
Low odour.
Use   All types of interior decoration where a decorative and hardwearing surface is important, e.g. corridors in schools, hospitals and public buildings; restaurants; clubs; food factories. Particularly suitable for use in occupied buildings because of its low odour.
Application

Medium or large set-ups give best results. Pressure should be about 20–30 psi (1.5–2 bars).

Drying
Method Coalescence.
Time Touch-dry in 4 hours; hard overnight.
Cleaning solvent Water or manufacturer's recommended cleanser.
Spreading capacity 14–20 m² per 5 litre.

(c) Brush applied, vinyl medium

Consists of two parts: (i) an undercoat which provides the ground colour, and (ii) the finishing coat which carries the coloured flakes.
Composition – undercoat
Pigment Titanium dioxide with extenders and coloured pigments.
Film-former Copolymer emulsion.
Composition – finishing coat
Pigment No pigment.
Film-former Coloured flakes are suspended in clear copolymer emulsion.
Properties
Non-toxic.
No objectionable odour.
Good adhesion.
Washable.
Abrasion-resistant.
Good flexibility.
Camouflages poor substrates.
Use For interior and exterior surfaces, particularly in public buildings where resistance to abrasion is of prime importance.
Application Undercoat is brushed or rolled. Finishing coat is brush applied. It may be necessary to roll the coating after to remove excess resin.
Drying Method Coalescence.
Time Undercoat, 2–4 hours. Finishing coat, touch-dry in 1 hour.
Cleaning solvent Water
Spreading capacity Undercoat – up to 70 m² per 5 litre. Finishing coat – up to 50 m² per 5 litre.

2.3.7 MASONRY PAINT

(a) Emulsion-based

A durable finish for outside walls (excluding timber).
Composition
Pigment Titanium white with extenders and coloured pigments.
Film-former Acrylic or vinyl copolymer resins.
Additives Mica, mineral quartz, granite and sand. One or more of these additives can be used to give reinforcement and texture to the film.
Thinner Water.
Properties
Alkali-resistant.
Good opacity.
Good filling properties.
Tough, durable and flexible.
Wide colour-range.
Quick and easy application.
Non-toxic.
Non-reversible coating.
Use For protection and decoration of new and old cement rendering, concrete, brickwork, asbestos sheeting, pebbledash, stucco and other types of masonry.
Application

Drying
Method Coalescence.
Time 1–4 hours.
Overcoating Can be overcoated as soon as the film is dry. Any type of decorative finish can be used over masonry paints as long as the surface is sound.
Cleaning solvent Water. It is very difficult to remove all fibrous material from brushes and rollers and they should not subsequently be used in normal finishing systems.
Spreading capacity 35–40 m² per 5 litre on smooth non-porous surfaces. Will vary according to the porosity and texture of the surface.

(b) Oil-based

A durable finish for outside use on all suitably primed or sealed surfaces.
Composition
Pigment Titanium white with extenders and coloured pigments.
Film-former Oil-modified resin or modified phenolic-coumarone resins.
Thinner White spirit
Additives Fungicides.
Properties
Good colour-range.
Smooth semi-gloss finish.
Tough and durable.

Non-reversible coating.

Use Interior, exterior protection and decoration of dry, primed plaster, brick, stucco, concrete, asbestos, sheeting, cement rendering, wood and metal.

Application

Drying Method Oxidation.

Time 4–8 hours.

Overcoating Overnight drying should be allowed. Can be overpainted with any kind of decorative paint.

Cleaning solvent White spirit.

Spreading capacity 70 m² per 5 litre on smooth non-porous surfaces. Will vary according to the porosity and texture of the surface.

Note Some masonry paints are based on synthetic rubber, pliolite resins.

2.3.8 CELLULOSE COATING

Quick-drying highly-flammable material, based on nitro-cellulose. Available as primer, sealer, filler and pigmented, clear or metallic finish.

Composition

Pigment Titanium white and a wide range of coloured pigments, extenders, slate powders and metallic powders.

Film-former Basically nitro-cellulose, but many types are modified with synthetic resins for greater flexibility and improved resistance.

Thinner Mixture of solvents according to types of cellulose and their uses. Include hydrocarbons, esters, ethers and ketones.

Properties

Very fast-drying.

Very hard and abrasion-resistant.

Water-resistant.

Reversible coating.

Use

 (i) Industrial and car refinishing.

(ii) Aluminium and bronze paints.

Application

Drying

Method Evaporation (some polymerisation with modified types).

Time Touch-dry in 10–15 minutes.

Overcoating 30–60 minutes depending on type, temperature and humidity.

Cleaning solvent Manufacturer's special thinner.

Spreading capacity For finishing coats these materials are sometimes thinned 50/50 with solvents. Therefore the spreading capacity covers a wide range, according to use.

Note Cellulose finishes provide very thin films; they are often brittle, and can sometimes crack, especially on timber surfaces.

Being reversible coatings, they are almost impossible to apply by brush except on very small areas.

The solvent odour from cellulose materials is strong and good ventilation is essential when spraying them.

These materials are highly flammable (see Part 6).

2.3.9 MOISTURE VAPOUR PERMEABLE (MVP) OR MICROPOROUS COATINGS

A paint system which allows moisture vapour to permeate through the dried film, but will not allow liquid water through. Such coatings allow the moisture content of timber to escape without causing the paint to blister or flake. Two types are available:

(a) Spirit thinned

(b) Water thinned

(a) Spirit thinned MVP coatings

Composition

Pigment Titanium white and a range of light fast pigments to give a full colour range.

Film-former Oil modified alkyd resin.

Thinner White spirit.

Properties

Mid sheen to full gloss.

Good flow.

Excellent flexibility.

Excellent durability.

Non-reversible coating.

Low maintenance.

Use All interior and exterior finishes. Can be applied over any bare timber substrate, or directly onto similar coatings in sound condition.

Application

Drying
Method Oxidation.
Time Touch dry in 3–5 hours.
Overcoating Leave overnight to dry.
Cleaning solvent White spirit.
Spreading capacity 12–15 m² per litre.
Maintenance Wash, spot prime, make good, apply two coats MVP coating every 4–6 years according to exposure.

(b) Water thinned MVP coating
Composition
Pigment Titanium white and a range of light fast pigments to give a full colour range.
Film-former Acrylic resin.
Thinner Water.
Properties
Semi-gloss finish.
Good flow.
Excellent flexibility.
Excellent durability.
Non-reversible coating.
Low maintenance.
Use All interior and exterior finishes. Can be applied over any bare timber substrate, or directly onto similar coatings in sound condition.
Application

Drying
Method Coalescence.
Time Touch dry in 2–4 hours.
Overcoating 4 hours in good drying conditions.
Cleaning solvent Water.
Spreading capacity 12–15 m² per litre.
Maintenance Wash, spot prime, make good, apply two coats MVP coating every 4–6 years according to exposure.

2.3.10 LOW ODOUR EGGSHELL FINISH

A water-borne coating for interior use which dries to a soft semi-gloss or eggshell finish.
Composition
Pigment Titanium white and a range of light fast pigments to give a full colour range.
Film-former Acrylic resin.
Thinner Water.
Properties
Low odour.
Quick drying.
Soft sheen.
Non-reversible coating.
Non-toxic.
Requires no undercoat.
Use Decorative finish for all interior surfaces.
Application

Drying
Method Coalescence.
Time 1–2 hours.
Overcoating 2–4 hours.
Cleaning solvent Water.
Spreading capacity 60–80 m² per 5 litre.

2.4 SPECIAL PAINTS

2.4.1 ANTI-CONDENSATION PAINT

An oil-based paint which dries with a textured finish and helps to reduce condensation. Similar paints based on emulsion resins are also available.

Composition Flat oil paint, eggshell finish or emulsion paint into which is added cork flour or similar material having low thermal conductivity. The cork granules act as an insulator between a warm moist atmosphere and the cold surface (Fig. 2.5).

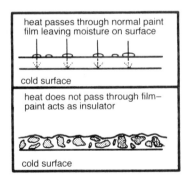

Fig. 2.5 *Anti-condensation paint*

Properties
Texture finish.
Reduces condensation.
Low sheen.
Retains its properties when overpainted.
Does not absorb moisture.
Non-reversible coating.
Use For interior ceilings and walls which are subjected to the effects of condensation. These paints are suitable for all surfaces but it should be noted that when applying the emulsion type, all bare metal should be well primed and undercoated with an oil paint before applying anti-condensation paint.
Application The paint can be sprayed or applied liberally by brush and finally hair-stip-

pled to ensure an even texture. Two coats are normally required for full effectiveness.
Drying
Method Oil-based, by oxidation; emulsion-based, by coalescence.
Time 8–12 hours according to thickness of coating, type of medium and conditions during application.
Overcoating After 16 hours. Further decoration can be carried out without impairing the properties.
Cleaning solvent For oil-based types: white spirit. For emulsion type: water.
Spreading capacity 25–30 m^2 per 5 litre. Maximum 20 m^2 per 5 litre for second coat.

2.4.2 ANTI-FUNGUS PAINT

Contains poisonous compounds which will kill fungus when it starts to feed on the paint film.
Composition Two kinds are in use:
(a) Those whose ingredients are specially chosen for their ability to resist or discourage fungus growth. They remain effective throughout the life of the paint film.
(b) Traditional paints which have fungicides added by the manufacturer on request. The anti-fungus properties do not always last the life of the paint film.
Properties In addition to the properties of the base paint it will also be mould-inhibiting and non-reversible coating.
Use
Where conditions are favourable for the growth of fungi, i.e. damp and humid conditions as found in bakeries, breweries, dairies and greenhouses. It is important that surfaces should be thoroughly prepared and sterilised before painting. Most leading manufacturers supply an antiseptic solution for this purpose. Extreme care should be taken to avoid contamination when using anti-fungus paints in places where food is being prepared.

Application

The addition of anti-fungus compounds makes the paint semi-toxic, therefore spray application is not recommended. Where there is no alternative to spray application, adequate ventilation should be provided and masks worn over mouth and nose, preferably with a separate air supply.
Overcoating Further applications must be of an anti-fungus type if such properties are still required.
Cleaning solvent Manufacturer's recommended solvent.
Note Tools, etc., used in preparation should be thoroughly sterilised to avoid spreading spores to other areas.

2.4.3 BITUMINOUS PAINT

A black paint used widely for the protection of steelwork and concrete.
Composition
Film-former Bitumens, which are available in different forms, blended together to produce the required type of finish. Heat radiation from the sun causes softening of the film and for this reason aluminium pigments may be added to give reflective properties; it is then known as aluminium bituminous paint.
Thinner Solvent naphtha or white spirit.
Properties
Chemical-resistant.
Moisture-resistant.
Low cost.
Gloss finish which is reduced by exposure to sunlight.
Reversible coating.
Use To provide a weather-resistant film on iron and steelwork, roof sheeting, storage tanks, concrete fabrications, bridges, piers, soil pipes and gutters.
Application

Drying
Method Evaporation.
Time 30–60 minutes.
Overcoating These paints are soluble in oil paint solvents and should be overpainted only with further bitumen coatings or be sealed to prevent bleeding through subsequent coats (see 3.4.7).
Cleaning solvent White spirit, paraffin or solvent naphtha. Avoid using the same brushes for other types of paint to avoid contamination.
Spreading capacity Varies with temperature and nature of surface at the time of application; average of 30–40 m^2 per 5 litre.
Note A limited range of coloured bitumen paints is available but these are more expensive and less resistant to moisture and chemicals.

2.4.4 CEMENT PAINT

Supplied in dry powder form and sold by weight rather than volume. The powder is mixed with water before use.
Composition White portland cement with extenders, waterproofing agents and accelerators. Alkali-resisting pigments are used to produce pastel shades.
Properties
Alkali-resistant.
Good filling properties.
Rough-textured matt surface.
Low cost.
Non-reversible coatings.
Use For the protection and decoration of brick, concrete, cement rendering, pebbledash, asbestos sheeting. Can also be used on similar interior surfaces in outbuildings, cellars, garages, farm buildings. Specially suitable for surfaces where damp and humid conditions would present problems for ordinary paints.
Note Not suitable for use on gypsum plasters.
Application

Fibre brushes should be used, as alkali destroys pure bristle. Must not be applied over other types of decorative paint. Porous surfaces should be damped down to control suction and the paint applied while the surface is still damp.

Drying
Method   By setting of the cement.
Time   2–4 hours according to the conditions.
Overcoating   Cement paints should be sealed with alkali-resistant primer/sealer before over-coating with ordinary decorative paints.
Cleaning solvent   Water.
Spreading capacity   Fairfaced brickwork and smooth surfaces, 8–15 m² per 5 kg; roughcast, pebbledash and tyrolean, 6–8 m² per 5 kg.
Note   Available in limited colour-range; poor flow and brushing properties.

2.4.5 CHLORINATED RUBBER PAINT

Chemical and water-resistant paint used for special purposes. Special primers are available for most types of surface. Thixotropic materials are available for heavy film-build.
Composition
Pigment   Titanium white and a limited range of chemical-resistant pigments, red lead and zinc chromate for primers.
Film-former   Chlorinated rubber.
Thinner   Manufacturers' recommended thinner.
Properties
Very good resistance to diluted alkalis.
Excellent water-resistance even when immersed.
Produces very flexible film.
Quick-drying.
Excellent resistance to acids and acid fumes.
Gives off chlorine gas when heated (limited fire-retardant properties).
Reversible coating.
Use
  (i) Applied to clean rust-free metal and wood, concrete, asbestos brickwork and insulation boards. Suitable for areas where special water or chemical resistance is required, such as plating shops, factories, laboratories and power stations.
 (ii) Swimming pools and water storage tanks.
(iii) Road-marking or line paint. When used for this purpose minute glass spheres, known as Ballotini spheres, are sometimes mixed with the paint to increase its light-reflectance. They may be added to the paint before brush application, but when spraying they may need to be dusted on to the

wet paint after application. Sand is also added to these paints to improve skid-resistance.
Application

Drying
Method   Evaporation.
Time   Touch-dry in 30 minutes.
Overcoating   8–10 hours by spray; 24 hours by brush. (Being a reversible coating, application by brush may sometimes be difficult, as the wet film softens previous coatings.)
Cleaning solvent   Manufacturer's recommended solvent.
Spreading capacity   45–55 m² per 5 litre; 10–15 m² per 5 litre for thixotropic types.
Note   These paints are not suitable for temperatures above 65°C, hot water above 43°C, surfaces in contact with drinking water or for immersion in solvents (the film will soften and swell).

2.4.6 DAMP-PROOFING COMPOUND

A waterproof, dark-coloured flexible coating, used as a barrier against dampness.
Composition   Prepared from bituminous materials, emulsified with an agent such as an alkaline soap. Sometimes pigmented with lime-resistant pigments and dyes.
Properties
Water-resistant.
Unaffected by humidity or temperature change.
Alkali-resistant.
Use
  (i) To hold back dampness on brick, plaster, cement, asbestos and most building materials. These compounds are not designed to cure damp but will help to check it for a limited time.
 (ii) Exterior stucco and concrete, particularly below ground level.
(iii) Sealing porous roof felt.
(iv) Non-reversible coating.

Application

Drying
Method   Evaporation.
Time   Within one hour.
Overcoating   Water-based acrylic and emulsion paints can be applied over them. A bitumen sealer should be applied before using oil-based paints. If the surface is to be papered, two coats of emulsion paint should be applied and allowed to dry thoroughly.
Cleaning solvent   Water.
Note   Brushes used for this material should not be used for any other type of paint.

2.4.7 EPOXY RESIN PAINT

Chemical-resistant primers and finishes available as two-pack materials, either pigmented or clear. Also known as cold-cured epoxy.
Composition
Pigment   Titanium white or metallic powders, with extenders and a wide range of coloured pigments.
Film-former   Epoxy resins.
Thinner   Manufacturers' recommended thinner.
Properties
Resistant to acids, alkalis, solvents and water.
Excellent adhesion to metals.
Abrasion-resistant.
Non-toxic.
Non-reversible coating.
Use
 (i) Where tough chemical-resistant surfaces are required e.g. in chemical plants, dairies, laundries, laboratories.
 (ii) Floor paints and other surfaces subject to abrasion.
(iii) Swimming pools or water storage tanks.
Application

Can cause health problems when sprayed. Has been banned in some countries.
Safety   Two-pack epoxies can cause health hazards if splashed in the eyes, on the skin or if spray mist is inhaled. Always read and observe the manufacturer's safety instructions.
Drying
Method   Polymerisation.
Time   2–4 hours; maximum hardness and chemical resistance after 7–10 days. Pot-life: day of mixing only.
Overcoating   14 hours.
Cleaning solvent   Manufacturer's recommended solvent.
Spreading capacity   55–65 m^2 per 5 litre.
Note   A single-pack epoxy paint based on epoxy esters is also available. It is thinned with white spirit and its chemical-, abrasion-and water-resistance is considerably inferior to that of the two-pack type.

2.4.8 FLAME-RETARDANT PAINT

Paint designed to reduce surface spread of flame on combustible surfaces. There are two main types:
(a) *Intumescent*, where the paint swells or froths when heated and produces a hard crust which acts as a barrier to prevent combustion.
(b) *Non-intumescent*, those containing substances which fuse together upon heating, to prevent combustion. They also give off a fire-retarding gas when heated.
Composition
(a) Chemical compounds which, when heated, expand rapidly to form a cellular skin approximately 30 mm thick over the surface. This acts both as a barrier to the flame and as a poor conductor of heat which helps to keep the substrate relatively cool.
(b) Paints pigmented with antimony oxide, with chlorinated rubber or chlorinated paraffin.
Properties
Greatly reduces the surface spread of flame on combustible substrates (see 3.4.1).
Easy to apply.
Available in a range of colours.
Non-reversible coating.
Use   To upgrade the classification of flammable building substrates in order to satisfy fire regulations in public buildings, exhibitions, film and television sets.

Intumescent paints are very useful for the protection of exposed steel girders and pillars in factories and warehouses where a fire would severely distort the steel supports, rendering the shell of the building unsafe.

Application

Drying
Method   Emulsion-type, by coalescence; oil-type, by oxidation.
Time   Emulsion-type, 2 hours; oil-type, 4–5 hours.
Overcoating   After overnight drying. Overcoating with orthodox paints greatly reduces the flame-retardant properties.
Cleaning solvent   Water or white spirit, depending on type.
Spreading capacity   30–35 m^2 per 5 litre.

2.4.9 MICACEOUS IRON OXIDE PAINT

A high-build material with outstanding protective properties for use on primed steelwork in exposed atmosphere.
Composition
Pigment   Micaceous iron oxide.
Film-former   Linseed oil phenolic resin medium.
Thinner   White spirit.
Properties
High build.
Good protection of sharp angles and edges.
Corrosion resistance.
Non-slip properties due to slight texture.
Non-toxic.
Chemical-resistant.
Low gloss.
Tough film.
Water-resistant.
Non-reversible.
Use   For protection of steel structures such as towers, bridges, marine installations, storage tanks and cranes. Excellent results when applied over metallic lead or red lead primers.

Application

Drying
Method   Oxidation.
Time   6–12 hours.
Overcoating   After 24 hours.
Cleaning solvent   White spirit.
Spreading capacity   35–45 m^2 per 5 litre.

2.4.10 FLUORESCENT PAINT

Brilliantly coloured paint which is highly visible even in poor light.
Composition
Pigment   Dyed urea formaldehyde resins dried and powdered.
Film-former   Non-yellowing medium.
Thinner   Varies with manufacturer and with type of paint.
Properties
Absorbs ultra-violet light and emits at greater intensity.
Translucent: poor opacity.
Use
(i)   Safety precautions: identification of hazards on sites.
(ii)   For display and advertising work where high visibility is required.
These paints are only efficient when applied over white grounds. They have poor durability and should be protected with a coat of clear lacquer.
Application

Also as a screen-printing ink.
Drying
Method   Depends upon the medium.
Time   2–4 hours according to type of medium used.
Overcoating   6–8 hours.

2.4.11 POLYURETHANE PAINT

Chemical-, water- and abrasion-resistant pigmented coating available as a two-pack material, requiring to be mixed shortly before use.

Composition

Pigment Titanium white and coloured pigments.

Film-former Polyurethane resins.

Thinner Manufacturer's special solvent.

Properties

Chemical-resistant.

Water-resistant.

Abrasion-resistant.

Good adhesion.

Heat-resistant.

Non-reversible coating.

Use

(i) Where tough chemical-resistant surfaces are required, e.g. in chemical plants, dairies, laundries and laboratories.

(ii) Floors and other surfaces subject to scratching and abrasion.

Application

Safety Two-pack polyurethanes can cause health hazards if splashed in the eyes, on the skin or if spray mist is inhaled. Always read and observe manufacturer's safety instructions.

Drying

Method Polymerisation.

Time Touch-dry in 3–5 hours. Maximum hardness and chemical resistance after 7–10 days. Pot-life: day of mixing only.

Overcoating Allow 24 hours between coats.

Cleaning solvent Manufacturer's special solvent.

Spreading capacity 55–65 m^2 per 5 litre.

Note A single-pack polyurethane paint based on urethane oils is also available. It is thinned with white spirit and its chemical-, abrasion- and water-resistance is considerably inferior to the two-pack type, although slightly better than alkyds.

2.4.12 TEXTURE PAINT (EXTERIOR)

These paints are designed to give texture and protection to smooth exterior concrete, cement renderings, asbestos cement, brick and exterior grade ply. The degree of texture varies considerably according to the type of pigment or aggregate employed in the coating. The texture is produced either (a) by manipulation of the wet film or (b) by particles of aggregate included in the paint.

Composition

Pigment (a) (b) Inert pigments and extenders similar to those used in interior texture paints (see 2.4.13).

Film-former (a) Emulsion; (b) Alkyd resin.

Aggregate (b) Fine granules of granite, stone, vermiculite, perlite or powdered rock. These are usually mixed into the paint, but in some preparations they are sprayed in the dry state on to previously applied wet film-former.

Thinner (a) Water; (b) White spirit or manufacturer's special solvent.

Properties

Good adhesion.

Weather-resistant.

Self-cleaning.

Thick coatings.

Good filling properties.

Flexible.

Non-reversible coating.

Use

(a) Used mainly for low traffic areas, e.g. claddings for domestic exteriors, ceilings in exterior corridors or subways.

(b) Much more robust and will therefore stand up to the abrasion and general wear encountered in public places, e.g. subways, underpasses, shopping precincts, multi-storey car parks, communal stairways, etc.

Application

(a)

Brush application is normally followed by stippling with a fine rubber stippler.

(b)

Where spray application is recommended, only suitable industrial equipment should be used as the abrasive nature of the aggregates damages standard equipment. Special spray gun set-ups are available based on hopper or pressure-feed systems; manufacturers of materials will give advice on request.
Covering capacity
Approx 25 m² per 5 litre.

2.4.13 RELIEF TEXTURE PAINT (INTERIOR)

Sometimes referred to as 'plastic paint', it forms an extra-thick surface coating which can be manipulated to give a range of relief textures. There are two main types available:

Type A
Supplied in powder form.
Composition
Pigment/aggregate China clay or whiting, exfoliated mica, mica and hemi-hydrate plaster.
Film-former gum arabic.
Thinner Water.
Properties Excellent adhesion.
Crack-resistant.
Flame retardant (class 0).
Does not encourage mould growth.
Reversible coating.
Use
 (i) Crack-resistant systems for plasterboard ceilings.
(ii) Texture finish to other smooth surfaces, e.g. plaster, cement rendering, concrete.
Application
Applied by brush to the surface and manipulated while still wet. Various tools are used to produce the required texture, e.g. stipplers, combs, sponges and palette knives. Some textures are laced with a plastic spatula to remove sharp edges and smooth out the high relief (see 1.1.12). Other effects can be achieved by direct application with spray or roller.

Drying
Method Evaporation.
Time In good conditions the surface will dry in 12 hours but initial setting takes only $\frac{1}{2}$ hour, after which time the coating cannot be reworked.
Overcoating To make the surface washable and also to improve its appearance it may be coated with either emulsion paint or with an oil-type primer/sealer followed by an eggshell finish. A further treatment can be scumbling and wiping.
Cleaning solvent Water.
Spreading capacity For average texture approx 0.5–0.8 kg/m².

Type B
Supplied as a paste ready for use.
Composition
Pigment/aggregate Titanium white, earth colours and coarse aggregates such as powdered rock.
Film-former Emulsions, sometimes blended with latex.
Thinner Water.
Properties
Excellent adhesion.
Flexible film – resists cracking.
Self-coloured.
Flame retardant (class 0).
Does not encourage mould growth.
Alkali-resistant.
Non-reversible coating.
Use
 (i) Heavy coatings applied to disguise most building surfaces including bricks and blocks.
(ii) Provides a wide range of decorative textures on virtually all common building surfaces.
(iii) Some types can be used on external surfaces.
Application Roller. Special rollers available for heavy applications. It can be manipulated while still wet to form various textures using stipplers, sponges, combs and plastic spatulas (see 1.1.12).
Drying
Method Coalescence.
Time 20–30 minutes to set. Touch-dry in 4 hours.
Dry in 24 hours.
Cleaning solvent Water.
Spreading capacity $5\frac{1}{2}$–10 m² per 25 kg.

2.4.14 SELF-CLEANING PAINT

Designed to chalk freely, thus enabling rain to wash away surface pigment along with dirt and grime.
Composition
Pigment Titanium anatase with extenders.
Film-former Medium to short oil length varnish.
Thinner White spirit.
These components are not normally used for exterior grade paints because of their chalking characteristics in exposed conditions. When used in self-cleaning paints they initially have some degree of gloss but this slowly disappears, leaving a constantly chalking film.
Properties
Controlled chalking.
Good opacity.
High build.
Non-reversible coating.
Use Surfaces which require to be kept clean but are not easily accessible, such as large wall surfaces where maximum light reflectance is essential.
Application

Drying
Method Oxidation.
Time 4–6 hours.
Overcoating If self-cleaning properties are to be retained, similar materials should be used for overcoating.
Cleaning solvent White spirit.
Spreading capacity 45–50 m^2 per 5 litre on smooth non-absorbent surface.

2.4.15 HEAT-RESISTANT PAINT

Remains protective and decorative when used on surfaces subject to heat.
Composition The components are carefully selected according to the temperature of the surface. White finishes will discolour at temperatures above 65°C and coloured pigmentation becomes impracticable after about 200°C. Aluminium powder is used for most finishes in the higher temperature range. Commonly in use are the following (see Fig. 2.6):

up to 90°C
Alkyd gloss finish Most makes will withstand temperatures of around 90°C, but white and very pale tints will discolour at 70°C and above.
up to 150°C
Heat-resisting acrylic paints Normally applied as a two-coat system on clean bare metal, omitting primer. May lift previous coatings. Surfaces should be cold on application and brought up to 100°C between coats and after finishing in order to effect proper curing.
up to 230°C
Aluminium paints Based on modified alkyd resin and aluminium powder. Suitable for inside or outside use. Two coats are normally applied to clean bare metal.
up to 260°C
Aluminium acrylic silicones Blends of acrylic and silicone resins with aluminium powder. Manufacturers' solvents are required for cleaning.
up to 260°C
Heat-resistant black Based on high grade bitumens and used on boiler casings, furnaces and exhaust stacks. Primers are not normally used. These paints are cheap compared with others in this range.

Fig. 2.6 *Heat-resistance of paints*

up to 540°C

Heat-resistant silicone aluminium Very expensive paints used on furnace fronts, heat exchangers and exhaust systems inside or outside. Silicone resins decompose at about 350°C but the aluminium becomes sintered to the metal and gives protection.

up to 500°C

Zinc dust/alkali silicates A two-pack material designed as a primer for steelwork and offering a tough, durable, heat-resistant base for systems which are subject to severe exposure and high temperatures (500°C).

Use Radiators and pipes on a normal domestic central heating system reach temperatures of about 65°C and most decorative paints can be used. Special heat-resistant paints are used for furnaces, exhausts, heat exchanges, boilers, steam pipes, chimney-stacks.

Choice of finish Some paints will discolour but remain protective and this may prove hazardous when the paint is used as an identification colour on pipelines. On a furnace or metal chimney-stack the main consideration would be protection rather than colour-retention.

2.5 CLEAR FINISHES AND WOOD TREATMENTS

2.5.1 WOOD PRESERVATIVES

Non-film-forming

There are three main types of non-film-forming timber preservatives which give protection against decay.

Composition

(a) *Coal-tar oils* These include creosote. This group is obtained from coal-tar distillates and is available in a number of grades. Heavy grades are suitable for pressure impregnation. Brushing and dipping grades are also available.

(b) *Organic solvent preservatives* Made by dissolving various chemicals in solvent, usually white spirit or solvent naphtha. After treatment the solvent evaporates leaving the preservative in the wood. Chemicals used include metallic naphthenates and pentachlorophenol.

(c) *Water-borne preservatives* Chemicals such as salts of copper and chromium dissolved in water.

Properties

Type (a) Low cost, strong odour, good resistance to insect and fungal attack, light or dark brown, non-corrosive, toxic to vegetation, difficult to paint over.

Type (b) Good penetration, good resistance to insect and fungal attack, non-corrosive, most types can be painted over when solvent has evaporated.

Type (c) Usually odourless, penetrates damp timber, clean to touch, can be decorated over, toxic to fungal spores.

Use

Type (a) Exterior timber, fencing and posts, telegraph poles, breakwaters, posts and all underground sections of timber, heavy structural timbers such as bridges, piers and railway sleepers.

Types (b) and (c) Internal and external timber. Both to protect structural timber such as beams and rafters also as a protection prior to painting on frames, doors, windows and cladding.

Application

The method of application is important.

Generally non-pressurised treatment results in poorer penetration.

(i) Pressure impregnation – most effective.
(ii) Steeping – timber soaked for days or weeks.
(iii) Dipping – timber submerged for up to 3 minutes.
(iv) Brush, spray or swab. Easiest, most common method for site work, but generally least effective.

Drying

Method Evaporation.

Time Depends on the rate of penetration, type and consistency of the preservative.

Overcoating Coal-tar preservatives should be overpainted only with similar materials. If orthodox paints are to be used the surface should be sealed with two coats of bitumen sealer or aluminium sealer. Other types of preservatives are suitable to receive paints and varnishes when dry.

Cleaning solvent For (a) and (b), white spirit. For (c), water.

Precautions When using these materials, masks and goggles should be worn and hands protected with barrier cream.

2.5.2 MVP STAIN/VARNISH

Durable finishes for natural woods in a range of translucent colours.

Composition

Pigment Translucent dyes and some earth pigments.

Film-former Oil modified resin with a microporous structure.

Thinner White spirit.

Properties

Allows wood to breathe.

Non-corrosive.

Enhances timber grain.

Good flow.

Excellent flexibility.

Aluminium powder

Types   Usually one standard grade available.

Properties

(i) Good silver colour.

(ii) Dulls quickly on exposure. Needs lacquering to maintain brilliance.

Use   Similar uses to bronze powders when a silver effect is required.

Gold sizes

The materials used to adhere metallic leaves and powders to a prepared surface. They are available in two forms as mordants or mediums depending on the method employed in gilding or bronzing.

Mordants

Applied to the surface and allowed to set to a tacky film before metallic leaf is applied to it, or powders are dusted on to it.

Three principal types are:

(i) *Japan gold size*   A quick drying, short oil varnish sometimes called signwriters size, or quick size. Available in a variety of grades depending on required gilding time, e.g. 1 hour; 2–4 hours; 4–6 hours. These times represent the approximate time the size takes to reach the ideal gilding tack.

Use   Small areas, generally for signwriting.

Properties   Provides minimum burnish.

(ii) *Oil gold sizes*   A range of sizes made from a blend of drying oils, resins and drying agents. They are available in many grades, e.g. 3 hour size; 12 hour size, and 24 hour size. Sometimes called French size or old oil size.

Use   Because of their slow setting time, they are most suitable for large areas of gilding, e.g. church finials, or extensive cornice and other relief moulding embellishment.

Properties   Allows more even film and produces a good burnish.

(iii) *Gloss paint*

Use   Equally useful on small or large areas of gilding.

Properties   Available in a wide range of colours; thick film can be applied which allow a good burnish; sets in 2–4 hours.

Mediums

Mixed with metallic powders to enable them to be applied as paints.

Japan gold size   Quick setting, suitable for small areas only.

Bronzing oil medium   Specially prepared varnishes which do not react with the acidic nature of the powders. Suitable for all areas of bronzing.

Bronzing cellulose medium   Specially prepared cellulose laquers for spraying on to properly prepared surfaces.

2.7.3 GILDING DEFECTS (Table 2.12)

TABLE 2.12 **Gilding defects**

Defect	Cause	Prevention
MISSES Patches where the leaf or powder has not adhered Show as dark or lustreless areas	Careless application	Apply size carefully. Tint size so that coating can be seen clearly as it is applied Apply size evenly to avoid thin areas drying faster than thicker areas Ensure surface is of even absorbency
WRINKLING Shrivelling of film so that it becomes contoured like the shell of a walnut	Size applied too thickly	Apply size evenly in normal thickness film Apply size in a more thin film
	Leaf applied too soon after application of the size. The leaf cuts off oxygen supply and the size remains soft underneath	Allow size to reach correct level of tackiness before applying leaf
	Pressure of thumb or cotton wool when applying leaf may wrinkle soft areas of size	Apply minimum of pressure when laying leaf
LACK OF BURNISH The metallic finish looks dull	Use of too thin size	Apply size in thicker film Use an oil gold size or gloss paint
	Bittiness of size caused by dust settling on size while setting-up	Screen sized areas with greased paper, aluminium foil or plastic
	Lack of polishing	Burnish well with cotton wool pad
VISIBLE JOINS OF LEAVES Edges of all leaves clearly seen in finish	Insufficient burnishing	Polish well with a cotton wool pad
	Use of hard metals which do not lose their leaf edges	Only use top quality gold leaf
PATCHINESS Uneven sheen or colour of a bronzing finish	Careless application	Apply size in even film Apply mixed paint evenly Stir and strain mixed paint thoroughly to ensure even distribution of metallic powder
RAGGED EDGE Edges of gilded work appear ragged	Excess chalk on surface during sizing	Dust off chalk with a soft brush

2.8 STENCIL AND MASKING MATERIALS

2.8.1 STENCIL-PLATE MATERIALS

Materials from which stencil-plates or templates are made are usually obtained ready prepared. A limited range can be made as required from commonly available materials. However, each should possess the following properties:
  (i) Easily cut with a sharp knife to produce a clean, smooth edge.
 (ii) Non-absorbent to paint and cleaning solvents used in the stencilling process.
(iii) Resist damage during use, and when cleaning.
 (iv) Remain flat during use, and during storage.

Types
Self prepared stencil paper Good quality drawing paper coated both sides with a mixture of linseed oil and liquid driers. After suspending for 2–3 days to allow thorough drying, it is ready for use. If a quicker drying coating is required, shellac knotting can be used to coat the paper.
Properties
Inexpensive.
Easy to cut.
Easily damaged, especially whilst cleaning.
Use
 (i) Small stencil designs
 (ii) Projects with few repeated applications and cleaning processes.

Ready prepared stencil paper Good quality card of varying thickness, impregnated with a drying oil. Available in A2 sized sheets, and either brown or yellow in colour.
Properties
Better resistance to damage.
Thicker varieties are difficult to cut.
More expensive than self prepared stencil paper.
Use
  (i) Small and larger stencil designs.
 (ii) Multi-plate stencils.
(iii) Projects with a medium number of repeated applications and cleaning processes.

Plastic stencil sheet Transparent acetate sheet containing a plasticiser to prevent splitting at the intersections of cuts.

Properties
Excellent resistance to damage.
Easy to cut.
Transparency allows easy and accurate alignment. Expensive.
Use
  (i) Small and larger stencil designs.
 (ii) Small sections of designs requiring accurate registration.
(iii) Multi-plate designs.
 (iv) Projects with many repeated applications and cleaning processes.

Metal stencils Ready prepared designs cut from thin metal sheeting. Also available are individual letters and numerals which can be linked together to produce simple signs.
Properties
Resists unlimited applications and washing processes.
Easy to store without damage.
Expensive.
Limited range of designs and sizes.
Usually single plate designs only.
Use
  (i) Small repeating patterns.
 (ii) Furniture decoration.
(iii) Friezes and borders.
 (iv) Simple signs.

Liquid masking or strippable coating
A water based film-former with limited surface adhesion when dry. After brush or spray application, and thorough drying, the stencil design is cut directly into the dry film. Sections of the film are easily removed, and paint applied. When the job is completed, the remaining film is removed, leaving the design painted on the surface.
Properties
Easy to hand cut.
Can be cut with knives set in compasses and trammels.
Will not lift from the surface whilst applying the stencilling paint.
Produces very clean edges to the finished work.

Use
 (i) One-off designs requiring clean-edged definition.
 (ii) Complex multi-colour projects.
 (iii) Work requiring a high level of accuracy.
 (iv) Geometric designs.

2.8.2 OTHER STENCIL MATERIALS

Masking tape
Rolls of self adhesive tape available in various widths from 6 mm to 300 mm wide. Made from paper or plastic film and possessing either standard tack or low tack adhesive qualities.
Use
 (i) In association with paper to mask areas from overspray when spray painting.
 (ii) Dividing one painted area from another in a clean line.

 (iii) Adhering drawings, pounces and stencil plates to the prepared surface.
 (iv) Low tack tapes used for similar purposes on surfaces which may be damaged by the standard tack variety.

Lining tape
Rolls of low tack plastic adhesive tape, which after application allow a central section to be removed, exposing the surface in a fine parallel line. Paint can be applied, and when tacky, the remaining tape removed to leave a fine painted line on the surface.
Use Painting fine lines on sign boards, panels and vehicles.

Liquid adhesive
Liquid adhesive in an aerosol spray container.
Use Sprayed on the back of stencil-plates and templates to hold them in position whilst applying paint by brush or spray.

Plate 1   *Bittiness*

Plate 2   *'Orange peel'*

Plate 3   *Rain-spotting or cratering*

Plate 4   *Runs, sagging or curtaining*

Plate 5   *Ropiness*

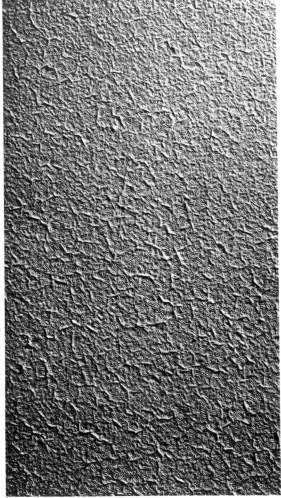

Plate 6   *Wrinkling, shrivelling
or rivelling*

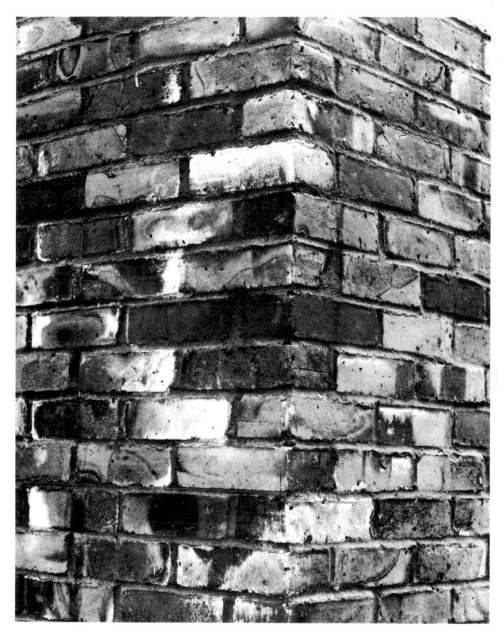

Plate 7   *Efflorescence on brickwork*

Plate 8   *Mould*

Plate 9   *Dry rot*

Plate 10   *Wet rot*

Plate 11   *Blisters on door panel*

Plate 12   *Cissing*

Plate 13   *Cracking or crazing*

Plate 14   *Flaking or peeling*

PART THREE

Preparation of Surfaces for Paint and for Clear Finishes

3.1 Abrasives   129

3.1.1 Coated abrasives   129
3.1.2 Abrasive powders and other abrasives   131

3.2 Making good   132

3.2.1 Stopper   132
3.2.2 Filler   132

3.3 Preparation of previously painted surfaces   135

3.3.1 Unsound surfaces   135
3.3.2 Sound surfaces   136

3.4 Preparation of new, stripped or untreated surfaces   138

3.4.1 Properties of building materials   138
3.4.2 Very absorbent surfaces   140

3.4.3 Absorbent surfaces   141
3.4.4 Non-absorbent surfaces   143
3.4.5 Chemically active surfaces   143
3.4.6 Corrodible surfaces   145
3.4.7 Bleeding and non-drying surfaces   148

3.5 Natural wood finishes   149

3.5.1 Preparation of natural wood surfaces   149
3.5.2 Finishes   149

3.6 Surface and paint film defects   151

3.7 Preparation of surfaces for graining, marbling and surface texturing   156

3.8 Preparation of surfaces for gilding and bronzing   157

3.1 ABRASIVES

Abrasives may be classified broadly into two groups: coated abrasives and abrasive powders.

3.1.1 COATED ABRASIVES (Table 3.1)

Natural or synthetic abrasive materials crushed, graded and stuck to paper or cloth with either a water-soluble or a waterproof adhesive. Used dry or with water, depending on type.

Paper

Weights of paper range between grades A to D Grade 'A': lightweight, flexible. Used to support finer grades of abrasive particles, for hand abrading. Grade 'D': heavyweight, rigid. Used to support coarser grades of abrasive particles, for hand and mechanical abrading.

Note Aluminium oxide abrasive paper can be obtained wrapped around and adhered to a rectangular block of synthetic sponge. They are used with water and available in fine, medium and coarse grades.

TABLE 3.1 Properties of coated abrasives

Abrasive	Manufacture	Grades	Use/Properties
EMERY CLOTH OR PAPER	Natural emery affixed to stout paper or cloth with a water-soluble adhesive	Coarse Medium Fine	Used dry or with a spirit lubricant for etching and degreasing metals before painting Very hard cutting edge Expensive
GARNET CABINET PAPER	Natural garnet 'open-coated' to grade D paper with a water-soluble adhesive	40 60 80 100	Used dry for the hand or mechanical abrading of wood before painting or varnishing Very sharp cutting edge
GARNET FINISHING PAPER	Natural garnet 'open-coated' to grade A paper with a water-soluble adhesive	150 180 220 240	Used dry for the 'flatting-down' between coats of varnish or french polish. Largely superseded by aluminium oxide papers
PRODUCTION PAPERS	Synthetic aluminium oxide 'open-coated' to grade D paper with a water-soluble adhesive	40 60 80 100 120	Used dry for heavy abrading operations by hand or by mechanical sanding tools Tough, very sharp cutting edge
ALUMINIUM OXIDE PAPER	Synthetic aluminium oxide open-coated to grade A paper with a water-soluble adhesive	150 180 220 240	Due to its long life and controlled cut it is superseding the use of garnet finishing papers

TABLE 3.1 **Properties of coated abrasives** – *continued*

Abrasive	Manufacture	Grades	Use/Properties
GLASSPAPER	Glass, natural flint or quartzite mineral affixed to grade D paper with a water-soluble adhesive	Strong (S) Coarse (C) Medium (M) Fine (F) Also 1, 0, 00 and Flour	Used dry for the hand or mechanical abrading of plaster, wood, boarding before painting, and denibbing between coats of paint Clogs, readily, therefore has a short life Tends to scratch the surface
SILICON CARBIDE PAPER (self-lubricating)	Synthetic silicon carbide 'open-coated' to grade A paper with a water-soluble adhesive. Paper additionally coated with zinc stearate which acts as a self-lubricant	220 240 320 400 500	Mainly used dry in the vehicle refinishing industry for abrading filling materials, 'flatting-down' between coats of paint and before painting Non-clogging
SILICON CARBIDE PAPER (wet or dry)	Synthetic silicon carbide affixed to a waterproofed grade A paper with a waterpoof adhesive	120 240 320 400 600	Used with water for abrading *all* types of previously painted surfaces before painting Superseded the use of pumice stone and other abrasive powders Tough and long-lasting if kept unclogged during use
TUNGSTEN CARBIDE	Synthetic mineral usually affixed to thin metal sheet	Fine Medium Coarse	Used for dry abrading of coarse painted surfaces, or new wood before priming Tough, long-lasting Extremely open-coated, so preventing clogging

Size

(i) Standard sheets of 280×230 mm.
(ii) Ready-cut rectangles or discs suitable for orbital or disc sanders.
(iii) Rolls suitable for belt sanders.

Grades of abrasive (Table 3.2)

Coated abrasives are graded by a number which corresponds to the size of abrasive particle used. For example, a coated abrasive graded 240 contains particles which will pass through a sieve containing 240 holes per square inch. Common grades range from 120 (coarse) to 600 (extremely fine).

TABLE 3.2 **Comparison of grades of abrasive**

	Fine			Medium			Coarse		
GLASSPAPER	00	0	1	$1\frac{1}{2}$	F2	M2	S2	$2\frac{1}{2}$	3
ALUMINIUM OXIDE	180		150	120	100	50	60	50	40
SILICON CARBIDE	600		400	280	240	180	150	120	100

Open-coated abrasives

Many coated abrasives tend to 'clog up' when used dry. Manufacturers overcome this problem by supplying a paper sparsely coated with abrasive particles which is ideal for mechanical sanding.

3.1.2 ABRASIVE POWDERS AND OTHER ABRASIVES (Table 3.3)

Usually naturally-occuring materials, crushed, graded and either (a) used with water and a felt block; (b) pressed into a block and used with water; or (c) made into a paste ready for use.

TABLE 3.3 **Properties of abrasive powders and other abrasives**

Abrasive	Origin or manufacture	Use/properties
PUMICE OR SODA BLOCK	Pumice powder and soda or detergent pressed into a block	Used with water for the initial cutting-down of very coarse paintwork before painting or paperhanging Liable to scratch severely.
WIRE OR STEEL WOOL	Fine tangled strands of steel or stainless steel wire	Coarse, medium and fine grades, for (i) degreasing and etching metal and plastics (ii) scouring timber after stripping with liquid paint remover (iii) flatting varnished or lacquered surfaces to produce a matt finish (Advisable to wear protective gloves)
SHOT OR GRIT	Abrasives particles obtained from many sources (see 1.2.3)	High-pressure abrasive cleaning (see 1.2.3)
RUBBING AND POLISHING COMPOUND	Mild abrasive powders mixed to a paste ready for use (water-soluble)	Used for rubbing and polishing synthetic and nitro-cellulose finishes

3.2 MAKING GOOD

The materials used in the process of repairing and surfacing before and during painting may be classified as stoppers and fillers.

3.2.1 STOPPER (Table 3.4)

A stiff material, which dries with the minimum of shrinkage, used for making good large holes and cracks.

Fig. 3.1   *Filling joints*

3.2.2 FILLER (Table 3.5)

A smooth paste used for filling slight surface imperfections.

Caulking

Caulking is a process of covering joints in dry lining. *Joints* (Figs 3.1 and 3.2). After the joints between boards or at angles have been filled, paper tape is sandwiched between a thin layer of filler. This work is carried out with a wide spatula or caulking tool.

Fig. 3.2   *Caulking process*

TABLE 3.4 **Properties of stoppers**

Stopper	*Use/properties*
HARD STOPPER	(i) A mixture of white lead, whiting and gold size (ii) Proprietary brands of oil stopper Traditional stopper for wood with good adhesion, very little shrinkage and some flexibility
LINSEED OIL PUTTY	Commonly used on wood for stopping holes and fixing glass Less expensive than hard stopper Becomes brittle with age
PLASTER	Stopper for gypsum and lime plasters Quick setting and drying Rake out and undercut for good adhesion For deep holes sand is added to the plaster to reduce shrinkage and cracking
SAND AND CEMENT MIX	4 parts sand to 1 of cement Used for (i) deep holes in plaster (ii) making good external rendering and brickwork
MASTIC	Very flexible material based on oil, latex or resin Mainly used around external door and window frames Usually applied with a caulking gun
CELLULOSE STOPPER	Used extensively in the vehicle refinishing industry under nitro-cellulose finish Quick drying
PLASTIC WOOD	Mixture of wood flour and resin Used under clear wood finishes, ready-mixed or 2-pack Quick setting Expensive
METAL STOPPER	Mixture of aluminium or iron powders and resin Used for stopping and re-building surfaces and arrises Usually 2-pack Quick setting Extremely hard Expensive
TINTED STOPPER	Water borne Used under clear wood finishes Available in a variety of colours to match stained wood
WATERPROOF SEALANT	Very flexible; based on latex or silicone resin Waterproof Tinted Filling cracks around baths and basins
TWO-PACK STOPPER	Very hard Takes screws and fixings Previous paint coatings should be removed Minimum shrinkage Rapid setting (15 minutes) Can be used in all clean, sound substrates

TABLE 3.5 Properties of fillers

Filler	Use/properties
PLASTER-BASED ('ALABASTINE' TYPE)	Used on plaster and for filling grain in timber Very absorbent when dry
PLASTER/CELLULOSE-BASED ('POLYFILLA' TYPE)	Used on plaster and for filling grain in timber Good adhesion, slight shrinkage on drying Absorbent when dry
PLASTER/VINYL-BASED ('INSTANT POLYFILLA' TYPE)	For use on internal surfaces Supplied ready-mixed Very smooth paste Shrinks on drying
CEMENT/VINYL-BASED ('EXTERIOR POLYFILLA' TYPE)	Used for external surfaces Excellent adhesion and water-resistance Brittle and difficult to abrade
SYNTHETIC EMULSION-BASED GRAIN-FILLERS ('FINE-SURFACE POLYFILLA' TYPE)	For internal and external surfaces Supplied ready-mixed Very smooth paste Can be worked to a fine feather edge Quick setting
OIL-BASED PASTE FILLER (SPACHTEL TYPE)	For internal or external painted surfaces Ready-mixed Low absorption when wet Slow drying Shrinks on drying
BRUSH OR SPRAY FILLER	(i) Ready-prepared with oil, synthetic or cellulose medium (ii) Oil-based, thinned with white spirit Several coats brushed or sprayed to flat or curved internal or external surfaces Low absorption
OIL-BASED RUBBING OR GRAIN FILLER	For internal and external use Supplied ready-mixed in natural and a range of colours Slow drying Applied by coarse cloth
ALL-PURPOSE FILLER	Available in powder form or ready mixed Suitable for interior and exterior surfaces Waterproof Flexible Can be used on most materials Good adhesion to metal, wood, masonry Very hard when set

3.3 PREPARATION OF PREVIOUSLY PAINTED SURFACES

3.3.1 UNSOUND SURFACES

Surfaces which are flaking, peeling, blistering, cracking, coated with size-bound distemper or otherwise unsound should be stripped back to bare surface by one of the following methods, and then treated as new surfaces.

Washing off

Surfaces, especially ceilings in older buildings, may be coated with size-bound distemper which must be regarded as an unsound surface.
 (i) Remove distemper by washing with water.
 (ii) Bind any residue of distemper with thinned alkali-resisting primer or proprietary binder/sealer (stabilising solution).

Burning-off

The fastest method of removing old defective paint and varnish coatings: they are softened by intense heat and scraped off with a stripping knife or shavehook.
Use   Mainly on timber surfaces.
Limitations   See Table 3.6.
Equipment   See 1.1.4.
Safety precautions (see also 6.2)
   (i) Do not overfill petrol and paraffin lamps.
  (ii) Allow ample warming-up time before pumping paraffin lamps.
 (iii) Supply adequate ventilation for inside work.
 (iv) Locate and have ready at hand suitable fire-fighting equipment before starting work (Fig. 3.3).
  (v) Protect floor from hot droppings with thin metal or non-flammable sheet.
 (vi) Avoid playing flame on any adjacent flammable surfaces.
(vii) Extinguish flaming droppings *immediately*.
(viii) Must not be used for removing old lead paint coatings.

Use of liquid paint-remover

Paint-removers soften old coatings of paint, varnish, and textured coatings allowing them to be removed with stripping-knife or shavehook. They are based on either solvent or alkali.
Use   Where removal by burning off is not practical or safe (see 6.4.2).
Limitations   See Table 3.6.
A. *Solvents*
(a) Highly flammable solvent, such as acetone plus wax to slow down evaporation and thicken the material.
(b) Non-flammable solvent, such as a chlorinated hydrocarbon plus methyl cellulose to slow down evaporation and thicken the material.
Properties
  (i) Softens the paint film and penetrates through to the substrate. The solvent evaporates, helping to lift the film.
 (ii) Does not damage timber, building boards, plaster, metal or stone surfaces.
(iii) Does not destroy bristle or hair.
(iv) Will remove most types of air-drying paint.
Safety precautions
  (i) Avoid causing sparks or naked flames (see 6.5).
 (ii) Provide adequate ventilation (see 6.6).
(iii) Protect linoleum and some plastics which may be damaged.
(iv) Protect hands with gloves.
B. *Alkalis*
(a) Proprietary brands based on caustic soda.
(b) Mixed on site by adding 0.5 kg caustic soda to 4 litres starch paste.

Fig. 3.3   *Fire-fighting equipment*

TABLE 3.6 **Comparison of effects of liquid paint-remover and heat-stripping**

Material treated	Liquid paint-remover	Heat-stripping
Coatings on large flat surfaces	Slow Messy Expensive Evaporation before the film is softened	Satisfactory
Coatings on intricately carved detail	Satisfactory	Disfiguration of detail by chipping with shavehook
Coatings adjacent to or on glass	Satisfactory	Cracking of the glass by heat
Coatings adjacent to or on flammable surfaces	Satisfactory	Ignition or melting
Coatings to be replaced by clear finish	Satisfactory	Disfiguration of the surface by scorching
Coatings on plaster and cement renderings	Satisfactory	Cracking and 'blowing' of the plaster from its backing
Coatings on thin metal sheet	Satisfactory	Buckling of the metal by heat (or melting of lead)

Properties
   (i) Saponifies the drying oils in the paint film.
  (ii) Destroys bristle and hair brushes.
 (iii) Difficult to remove all traces of alkali after use, especially on absorbent surfaces.
 (iv) Tends to lift the grain of timber and may lift veneers.
  (v) Cheap and non-flammable.
 (vi) Particularly effective used as a dip.
(vii) Corrosive to non-ferrous metals, especially aluminium.

Safety precautions
  (i) Wear rubber apron, gloves and goggles to prevent damage by accidental splashing.
 (ii) Protect surrounding areas, particularly linoleum and fabrics.

Mechanical abrading (see 1.2.3)
Most types of coating can be removed by the use of disc sanders or belt sanders.
Safety Respirators should be worn to prevent dust being inhaled.

3.3.2 SOUND SURFACES

Some surfaces may be firm and sound but contain a few areas where the coating is flaking as a result of corrosion or dampness. The procedure then is to scrape off all the loose paint film to a hard edge and to expose the surface if possible. The bare areas should be cleaned, abraded and primed (spot-priming). When the primer is dry the surface should receive the following treatment:

  (i) Wash and rinse to remove all traces of grease and grime which, if left, would reduce the adhesion of the undercoat. A variety of washing materials is available (see Table 3.7); when mixed with water. Ready for use, these are often called 'pickle', and are usually applied and scrubbed on with an old brush.
 (ii) Abrade to provide a smooth surface and a 'key' for the undercoat, again to increase adhesion. Use the abrasive (Table 3.1) on a

TABLE 3.7 **Types of washing material**

Material	Advantages	Disadvantages
SUGAR SOAP	Inexpensive Etches the surface	Softens paint film if mixed and used too strong Residue left on surface may slow down or prevent drying of paint applied over
WASHING SODA	Inexpensive Does not froth-up	Softens paint if mixed and used too strong (1 handful to 1 litre boiling water is usual strength for normal washing) Difficult to rinse off all traces Residue may affect drying of subsequent coats
DETERGENT POWDER	Removes grease very efficiently	Softens paint if mixed and used too strong Expensive Difficult to remove because of its frothing action
SOAP POWDER	Milder than detergent, does not soften paints	Expensive Difficult to remove froth and 'greasy' film Residue may affect drying of subsequent coatings
EMULSIFYING LIQUIDS	Excellent degreasing properties, particularly in removing thick grease and oils from absorbent surfaces	Expensive

felt, rubber or wooden block to maintain even pressure and a flat surface. Wet-abrading may be combined with the washing process.

Note For health reasons, do not dry-abrade painted surfaces which may contain lead (see 6.4).

(iii) 'Make good' or stop and fill all holes, cracks and other irregularities in the surface.

3.4 PREPARATION OF NEW, STRIPPED OR UNTREATED SURFACES

3.4.1 PROPERTIES OF BUILDING MATERIALS (Table 3.8)

All building materials can be classified as either (a) *porous materials* which have air voids in their composition, formed by liquid, usually water, evaporating or chemically combining with the material during its manufacture or formation (Fig. 3.4); or (b) *non-porous materials* which have no air voids in their composition, being maufactured or formed under conditions of extreme heat or pressure and allowed to cool; this drives off all water content, leaving no air voids (Fig. 3.5).

Fig. 3.4   *Porous material*

Fig. 3.5   *Non-porous material*

Materials which are commonly painted can for the purpose of surface preparation be further classified into six groups according to their physical and chemical properties.

(a) *Very absorbent* materials have air voids linked by capillary tubes, readily allowing the passage of liquids (Fig. 3.6).

Fig. 3.6   *Very absorbent material*

(b) *Absorbent* materials have air voids partially separated by impervious material such as resin. They are reluctant to allow the passage of liquids (Fig. 3.7).

Fig. 3.7   *Absorbent material*

(c) *Non-absorbent* materials do not allow liquids to enter or pass through them, because either they are non-porous, or they are porous but have air voids completely surrounded by impervious material, so that the movement of water is prevented (Fig. 3.8). (Also known as impervious materials.)

Fig. 3.8   *Non-absorbent material*

(d) *Chemically active* materials are alkaline and promote a chemical reaction with some drying oils in paints. The most common alkali found in building materials is lime (Fig. 3.9).

Fig. 3.9   *Chemically active material*

TABLE 3.8   **Properties of building materials**

Materials	Porous	Non-porous	Very absorbent	Absorbent	Non-absorbent	Chemically active	Corrodible	Non-drying or bleeding	Flame-spread classification
Glass reinforced cement board	X		X			X			0
Bitumen and coal-tar coated surfaces								X	1–4
Blockboard and plywood	X			X					3
Brickwork	X		X			X			0
Building blocks	X		X			X			0
Cement rendering	X		X			X			0
Chipboard	X			X					3
Concrete	X		X			X			0
Expanded polystyrene	X				X				4
Fabrics	X		X						4
Ferrous metals		X					X		0
Fire-proofed boards	X		X	X		X			1
Glass		X			X				0
Glass-reinforced plastics		X			X				3
Glazed tiles		X			X				0
Gypsum plaster	X		X			X			0
Hardboards and MDF	X			X					3
Hardwoods	X			X					2
Insulating board	X		X						4
Lime plaster	X		X			X			0
Non-ferrous metals		X			X		X		0
Oil-contaminated surfaces								X	1–4
Paper	X		X						4
Plasterboard	X		X						1

TABLE 3.8   **Properties of building materials** – *continued*

Materials	Porous	Non-porous	Very absorbent	Absorbent	Non-absorbent	Chemically active	Corrodible	Non-drying or bleeding	Flame-spread classification
Plastics		X			X				4
Resinous woods	X			X				X	3
Softwoods	X			X					3
Stonework			X		X				0
Wood wool slabs	X		X		X				1–0

(e) *Corrodible* metals will rust under certain conditions (Fig. 3.10).

Fig. 3.10   *Corrodible material*

(f) *Non-drying or bleeding* materials (i) contain or are contaminated with compounds which retard or prevent the drying of paints applied to them; (ii) will 'bleed' through and discolour paints applied to them (Fig. 3.11).

Fig. 3.11   *Non-drying or 'bleeding' material*

Flame-spread classification
All the above materials can be classified according to their flame-spread properties.

The Building Regulations specify minimum requirements in terms of flame-spread classifications for building materials. BS 476: Part 7: 1997 lays down the method of assessing materials so that they can be given a 'class' of flame-spread. The method is dependent on the distance that a flame will travel across a sample of material of given size. The classes of flame spread are:
Class 0   Non-combustible surfaces
Class 1   Surfaces of *very low* flame-spread
Class 2   Surfaces of *low* flame-spread
Class 3   Surfaces of *medium* flame-spread
Class 4   Surfaces of *rapid* flame-spread.

3.4.2 VERY ABSORBENT SURFACES

Insulating board
Wood fibres and resin lightly compressed to form a rigid, porous, lightweight building board. Used extensively for the internal cladding of ceilings and walls. Excellent sound-and heat-insulating properties. Available:
(a) untreated
(b) ready-primed
(c) fire-proofed (see 3.4.5)
(d) bitumen-impregnated (see 3.4.7).
Preparation
 (i) Ensure surface is dry and free from dust and grease.
 (ii) Punch nails below the surface and spot-prime.
(iii) Caulk joints and nail-holes (see 3.2).
(iv) *Do not* abrade, as this will cause surface damage.

Primer
(a) Low-pigmented oil scaler to penetrate and reduce absorption.
(b) Thinned emulsion paint.
(c) Acrylic primer/undercoat.
(d) Hardboard primer.
(e) Glue size solution.
Note Water-borne primers tend to raise the surface fibres of the board and corrode nail heads.

Acoustic board (Fig. 3.12)
Sheets or tiles of insulating board with a variety of holes or slots covering the surface. The surface texture increases the surface area, so improving the sound-insulating properties of the board.

Fig. 3.12 *Acoustic board*

Types, Preparation, Primer, as insulating board.
Note Short-pile paint roller will prevent the textured surface being filled with paint, which would reduce the acoustic properties of the board.

Paper and fabrics
With the possible exception of wallpaper, hessian and canvas, paper and fabrics are not usually painted.
Preparation Ensure that the surface is dry and free from dust.
Primer
(a) Glue size solution.
(b) Thinned emulsion paint.
(c) Acrylic primer/undercoat.
Oil-based primers must not be used because they will either rot or discolour the vegetable fibres of the material.

Fig. 3.13 *Plasterboard*

Plasterboard (Fig. 3.13)
Building board of hard gypsum plaster sand-wiched between two sheets of stout paper or card. Two usable sides: grey for plastering and cream for decorating. Used as dry lining.
Preparation, Primer, as insulating board.

3.4.3 ABSORBENT SURFACES

Timber

Moisture content
Timber contains moisture in varying amounts, and it is the drying out and absorption of water which causes wood to expand and contract, so damaging paint films. Before painting wood, ensure that moisture content does not exceed 18 per cent. Use a moisture meter to check this (Fig. 3.14). Painting timber having over 20 per cent moisture content may result in blistering and flaking (see 3.6).

Fig. 3.14 *Moisture meter*

Preservatives
Timbers with high moisture content can rot even under paint systems. To avoid this, special preservatives can be applied. Such treatment is recommended for exterior softwood joinery and cladding. Applied to all surfaces before fixing. It is common practice for self-coloured preservatives to be left unpainted.

Back-priming
Where wood is to be fixed to structures such as brickwork or plaster, the entire surface, parti-

end grain

Fig. 3.15 *End-grain requiring back-priming*

cularly end grain, should be primed before fixing to prevent any future penetration of water. Timber with over 20 per cent moisture content is open to attack by dry rot (Fig. 3.15).

Primer application

Good adhesion is obtained by brush application which ensures intimate contact between paint and substrate.

Shop-primed timber

Some timbers and building boards can be obtained already primed, but the priming paint and application are often inferior to 'on site' priming. If in doubt, dust off the surface and reprime with suitable primer.

Softwood

Softwood timbers are obtained from fast-growing trees. The main timber used in construction is from coniferous trees (evergreen, with needle leaves). Softwoods require protection by surface coatings. Common types are pine, cedar, spruce (whitewood), redwood (deal).

Preparation
  (i) Ensure that moisture content is low.
 (ii) Punch nails below surface.
(iii) Dry-abrade to a clean, smooth surface.
(iv) Prevent resin staining by coating knots and resinous parts with shellac knotting (see 3.6).
 (v) Apply preservatives when specified.

Primer
(a) Pink wood-primer.
(b) White wood-primer.
(c) Lead-free primer.
(d) Aluminium wood-primer.
(e) Acrylic primer/undercoat. Emulsion paint is unsatisfactory.
(f) Calcium plumbate primer (composite wood/metal components).

Hardwood

Hardwood timbers are obtained from slow-growing trees, deciduous (leaf-shedding), with broad leaves. They are not usually painted, but often given a clear finish to enhance the grain. Common types are ash, oak, beech, mahogany, sapele.

Preparation, as softwood.
Primer As softwood. Better penetration of the usually tighter grain may be obtained if the primer is thinned.

Fig. 3.16 *Blockboard*

Blockboard (Fig. 3.16)
Strips of softwood sandwiched and stuck between double ply boards. Rigid, smooth-faced building board 16–25 mm thick. No knots or resinous parts. Available: (a) standard, (b) veneered, usually clear-finished.
Preparation, Primer, as softwood.

Fig. 3.17 *Plywood*

Plywood (Fig. 3.17)
Thin veneers of wood stuck together with the grain running at right angles one to another. Rigid and of varying texture. Thickness 3–25 mm depending on number of veneers or plies. Grain often difficult to smooth by filling. No knots or resinous parts. Available:
(a) interior grade
(b) exterior grade
(c) marine (resin-bonded)
(d) veneered, usually clear-finished.
Preparation, Primer, as softwood.

Hardboard/Fibreboard

Wood fibres and resin firmly compressed at high temperature. Rigid, smooth-faced, even-textured building board. Available:
(a) standard, 3–12 mm thick. Smooth surface on one side, mesh texture on reverse.
(b) medium density fibreboard (MDF) from 12 mm thick. Less dense than standard hardboard.

(c) oil-tempered, with increased resistance to moisture. Very smooth surface.

(d) perforated: variety of holes and slots.

(e) prepared or ready-primed.

(f) fireproofed (see 3.4.5).

Preparation

 (i) Ensure that the surface is dry and free from dust and grease. Degrease with white spirit if necessary.

 (ii) Punch nails below the surface. Caulk joints and nail holes (see 3.2).

(iii) *Do not* abrade as this will cause surface damage.

(iv) Prime *before* making good, paying special attention to edges.

Primer

Internal: (a) Thinned emulsion paint; (b) acrylic primer/undercoat; (c) hardboard primer.

Note Pink primer is unsatisfactory.

External: (a) hardboard primer; (b) aluminium wood-primer.

Note Both thinned emulsion paint and pink primer are unsatisfactory. Extra durability can be obtained by back-priming.

Chipboard

Wood chips and resin firmly compressed at high temperature. Rigid, dense, even-textured building board 9-38 mm thick. Available: (a) standard, (b) medium, for furniture, (c) flooring quality. *Preparation, Primer,* as softwood.

3.4.4 NON-ABSORBENT SURFACES

Usually very dense, smooth and shiny, so offering very little adhesion for paints.

Glass, glazed tiles, glazed bricks

Preparation

 (i) *Do not* abrade as the surface may be damaged.

 (ii) *Glass* Degrease by washing with methylated spirit and whiting, allowing to dry to a powder and polishing with clean rag.

(iii) *Glazed tiles* Wash with strong sugar soap or detergent; rub joints between tiles with cellulose thinners.

(iv) Prime as soon as surface is dry. Do not allow further contamination, even with finger-prints.

Primer Must possess good adhesive properties:

thinned alkyd gloss paint or special tile bonding coat based on PVA or acrylic resin.

Plastics

A number of synthetic materials either in sheet form or capable of being moulded by a variety of methods. Usually self-coloured, therefore not often painted. Those most commonly encountered in painting are:

(a) PVC (polyvinyl chloride). Usually coloured white, grey or black. Used extensively for general soil, waste and water services.

(b) Glass-reinforced plastic (fibreglass). Plastic resin reinforced with layers of glass-fibre matting. Used for a large variety of moulded items. Often self-coloured.

(c) Melamine laminates. Decorative sheeting used for panelling, door-covering and table and counter tops.

(d) Acrylic (Perspex, Oroglas). Clear or coloured sheeting or formed units such as baths, storage tanks and signs.

Preparation

 (i) Degrease with detergent and rinse.

 (ii) Paint as soon as surface is dry with specified finishing paint. Do not allow further contamination.

Expanded polystyrene

Polymerised styrene resin expanded into an extremely lightweight, porous, insulative material. Used extensively for ceiling tiles and wall insulation sheets. Available:

(a) standard: highly flammable, the most common type.

(b) self-extinguishing: more dense than standard, slow to ignite so reducing surface spread of flame, provided it is fixed with a full film of adhesive and not 'spotted'.

Preparation Make good with plaster-based filler only, as oil filler may dissolve polystyrene.

Primer Emulsion paint or acrylic primer/undercoat.

3.4.5 CHEMICALLY ACTIVE SURFACES

Most chemically active surfaces are mixed with a considerable amount of water when being manufactured or used, and this water may take many months to dry out. If painting is essential before complete drying out (e.g. in new buildings), only permeable water-borne paints should be used.

Chemically active surfaces are those which contain *lime* or *cement*. They become active alkalis only when wet. Suspect surfaces can be checked for alkalinity by dampening with distilled water and applying red litmus paper. Alkali will change the colour of the paper to blue.

Saponification is the effect of alkali on most drying oils used in paints. The alkali changes the drying oil into a non-drying, water-soluble soap (see 2.6).

Efflorescence: most chemically active surfaces contain water-soluble salts which rise to the surface in solution and crystallise as the water dries out (Plate 7). Remove by dry brushing only, as washing will aggravate efflorescence.
Primer
Wet surfaces Emulsion paint or cement paint – permeable coatings which allow the water to dry out through the paint film without breakdown of the film.
Dry surfaces (a) Alkali-resisting primer/sealer (see 2.2.2); (b) thinned emulsion paint.

Gypsum plasters
Plasters produced by the burning of natural gypsum (calcium sulphate). Although they are almost lime-free, they are often applied to strongly alkaline surfaces which make them chemically active.
Classes according to BS 1191: Parts 1 and 2: 1994
Class A   Hemi-hydrate (plaster of Paris), used for fibrous plastering.
Class B   Retarded hemi-hydrate ('Thistle').
Class C   Anhydrous ('Sirapite').
Class D   Keene's cement.
Premixed: Class B plaster with the addition of exfoliated vermiculite and expanded perlite to produce a lightweight, insulative, fire-retardant plaster ('Carlite').
Preparation
  (i) Ensure that the moisture has dried out.
  (ii) Remove any efflorescence.
  (iii) Denib with stripping knife. *Do not* abrade as the surface will be damaged.
  (iv) Make good large holes and cracks with a similar plaster or plaster filler.
Primer (a) Thinned alkali-resisting primer. Thinning improves the primer's penetrative power.
(b) Thinned emulsion paint for emulsion paint finish.

Lime plaster
Plaster produced by the burning of natural limestone to make 'quicklime' which is then steeped in water to form 'lime putty'. Not often used today, but older buildings would certainly contain it. Hardens very slowly but often speeded-up by adding gypsum plaster (gauged lime plaster).
Preparation   As gypsum plaster.

Fig. 3.18   *Building block*

Building blocks (Fig. 3.18)
Pulverised fuel ash or clinker (burnt coke) and cement slurry. Lightweight, even-textured, loadbearing and non-loadbearing blocks. Used extensively for partition and cavity walls. Sometimes incorrectly called 'breeze blocks'.
Preparation
  (i) Ensure that the moisture has dried out.
  (ii) Clean and remove any efflorescence with a wire brush.
  (iii) Clean and sterilise any areas of mould growth or lichen (see 3.6).

Brickwork, stonework, concrete, cement rendering
Preparation   Treat as building blocks.

Glass reinforced cement
Glass fibres and cement slurry pressed to form sheets or cast in moulds. Dense, even-textured, fire-proof. Used extensively for rain-water goods, fire-proofing doors and walls and corrugated sheeting for roofs.
Preparation Dry-brush to remove dust. Abrading causes surface damage and is a health risk (see 6.6). Back and edges should be coated with bitumen or other water-resistant coating to prevent moisture penetrating and damaging the finish.

Woodwool slabs
Long tangled wood fibres and cement slurry

pressed into a lightweight building slab Good insulator, very coarse surface texture resembling straw. Commonly used in partition walls.
Preparation Dry-brush to remove dust.
Note Because of their open texture these surfaces are best coated by spray.

Fire-proofed boards

Building boards, especially insulating board and hardboard, which have been treated with water-soluble fire-retardant solutions to lower their class of flame spread (see 3.4.1).
Preparation According to type of board (see 3.4.2).
Primer Alkali-resisting primer/sealer only.

Do not use water-borne paints as they may promote efflorescence of the fire-retardant salts.

3.4.6 CORRODIBLE SURFACES

Ferrous metals

All metals which have iron in their composition are classed as ferrous metals. Cast iron, wrought iron, mild steel, high tensile steel and stainless steel are the most common examples. The primary difference between them is the amount of iron in the metal, which ranges from a very large amount (cast iron) to a very small amount (stainless steel).

When iron and steel are exposed to oxygen and water (usually in the atmosphere) they quickly corrode: reddish-brown deposit known as rust forms on the surface. It is the iron content which produces the rust deposit; metals with a large iron content corrode more readily than those with a small iron content. Rusting weakens the metal.

Chemistry of corrosion (Fig. 3.19) Corrosion is a chemical change brought about by an electro-chemical process called electrolysis. When metal is immersed in an electrolyte (water contaminated with compounds such as acids or salts), an electrolytic cell is set up. An electrical current will flow between certain areas of the metal surface known as anodes and cathodes. The current flows from the cathode through the metal to the anode and returns through the electrolyte, which is capable of conducting electricity. The current results in destructive corrosion of the anode area with the formation of rust. While the anode corrodes, the cathode is protected against corrosion.

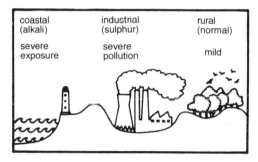

Fig. 3.20 *Atmosphere as the electrolyte*

When metals are exposed to the atmosphere, they are actually immersed in an electrolyte, since the air contains water, dilute acids and salts, especially in industrial and coastal areas. Corrosion is more severe in a strong electrolyte, which explains why metals rust more rapidly in industrial and coastal atmospheres (Fig. 3.20).
Millscale (Fig. 3.21) When ferrous metal is forged into flat plates, beams and girders, etc., it is worked at temperatures often exceeding 1000°C. While the metal is in this white-hot state it is in contact with the cool air and rapidly oxidises. This results in the formation of a thin,

Fig. 3.19 *The corrosion process*

Fig. 3.21 *Millscale*

flaky layer of black oxidised iron called mill-scale. The millscale is cathodic to the bare steel and when it is broken anodic areas are exposed.

Preventing corrosion To prevent rusting, ferrous metals can be completely enveloped in a thick, elastic paint system which must adhere firmly to the surface. This can only be accomplished by:

 (i) The complete removal of all rust and mill-scale.
 (ii) Use of a suitable rust-inhibitive primer.
 (iii) Use of a weather-resistant paint system of adequate thickness.

Fig. 3.22 *Removal of rust and millscale*

(i) *Surface preparation* (Fig. 3.22) Millscale and rust promote further corrosion and prevent the firm adhesion of paint films. Methods of removal, in order of effectiveness, are:

(a) Abrasive cleaning: a stream of abrasive particles is directed against the metal at a high velocity to dislodge all rust and millscale, leaving a clean, white metallic surface (see 1.2.3).

(b) Acid pickling: a factory method of treating metals by use of acids. One of the most common is the 'Footner process', which involves first dipping the metal in dilute sulphuric acid, then rinsing in hot water and finally dipping in dilute phosphoric acid. This leaves a black phosphate film on the surface, ideal for painting.

A less effective site process is to use an acid solution, mainly phosphoric, applied by brush during preparation of the surface.

(c) Flame-cleaning: a high temperature oxy-acetylene flame which causes rapid expansion of the metal, loosening the scale and dehydrating the rust, which is brushed off (see 1.2.2).

(d) Mechanical and hand methods: preferably carried out after weathering. Loosened rust and millscale is partially removed using chipping hammers, derusting pistols and wire brushes (see 1.1.3).

Removal of grease and ferrous sulphate The surfaces of metals are naturally greasy, especially around welded joints where oily fluxes are used during welding operations. To prevent paint films from flaking, the surface of the metal must be degreased before painting by washing with either naphtha, white spirit or an emulsifying agent. Also, washing with hot clean water removes ferrous sulphate salts which promote corrosion by attracting and absorbing water.

(ii) *Priming* Immediately after preparation, the metal should be primed with a special rust-inhibitive primer. Because of the effects of surface tension, arrises and bolt-heads should receive two coats. Where possible, apply by brush or roller. If spray application is necessary, airless spraying is superior to conventional methods. Factory-primed metal should be washed with detergent, rinsed, dried and reprimed after fabrication.

Rust-inhibitive primers The pigments used in these paints actually prevent the formation of rust even if the film is slightly damaged. They include red lead, metallic lead, calcium plumbate, zinc chromate, zinc phosphate and zinc dust.

Fig. 3.23 *The paint system*

(iii) *Paint system* (Fig. 3.23) The whole surface must be protected with a weather-resistant paint system which must not be less than 125 microns in thickness. This ensures that all peaks or high spots on the surface are adequately covered.

Note 125 microns represents approx 4–5 coats of conventional paint.

TABLE 3.9   **Non-ferrous metals – surface preparation and primers**

| Metal | Preparation | | Primer |
	Factory	On site	
Aluminium sheeting extrusions	Degrease in acid bath	Etch and degrease with emery cloth and white spirit	Etch primer Zinc chromate (No lead paints)
Galvanised iron (steel dipped in molten zinc)	Degrease in acid or alkali bath	(i) Allow to weather for several months *or* (ii) Degrease	(i) Zinc chromate (ii) Calcium plumbate or etch primer
Zinc sprayed sherardised	Denib with emery cloth and degrease	Denib with emery cloth and degrease	Etch primer followed by zinc chromate
Copper Brass Bronze	Degrease in solvent bath	Emery cloth and white spirit (do not weather or dry-abrade)	Etch primer
Lead	Not usual	Allow to weather for several months *or* Emery cloth and white spirit (do not dry-abrade for health reasons)	Etch primer

Non-ferrous metals

All metals which have no iron in their composition are classed as non-ferrous metals. Aluminium, zinc, copper, brass, bronze, lead, chromium, tin and cadmium are commonly used in buildings. Coated iron and steel may also be classed with non-ferrous metals e.g. galvanised iron.

Corrosion of non-ferrous metals   Non-ferrous metals, like ferrous metals, corrode by the chemical process of electrolysis, with the same influencing factors. The main difference is that, on the surfaces of non-ferrous metals, firmly-adhering corrosion deposits are formed which inhibit and slow down further corrosion. Corrosion can be accelerated by contact with dissimilar metals, acid or alkaline materials.

Surface preparation   Generally speaking, the surface must be dry and free from grease, dirt, corrosion deposits and loose paint. Methods of pretreatment are shown in Table 3.9.

Sacrificial coatings

One method of reducing corrosion is to apply a coating containing a metal which corrodes more rapidly than the substrate metal it is in contact with.

Metals are classified on their activity in the galvanic series (Table 3.10). Metals at the top of the series are more active than those lower down. If a coating containing a metal is applied to a substrate which is lower on the galvanic series, the metal in the coating will corrode rapidly, so 'sacrificing' itself for the substrate metal.

Galvanizing is a zinc coating and when applied to iron it is sacrificed and protects the iron from corrosion. Alternatively, if an aluminium surface is coated with a lead based coating, the aluminium will corrode, which is the reason why lead based primers must not be used on aluminium or zinc.

TABLE 3.10   **Galvanic series**

Aluminium	(unstable or active metals)
Chromium	
Zinc	
Iron	
Lead	
Copper	
Silver	
Gold	(stable or noble metals)

3.4.7 BLEEDING AND NON-DRYING SURFACES

(a) Surfaces coated or contaminated with bitumen, coal-tar oils, nicotine or mineral oil and grease.
(b) Surfaces or paints containing certain lake pigments or dyestuffs.

When a surface contains material which is soluble in the medium of the paint applied to it, the result may be discolouration by bleeding (see 3.6) or possible slow drying or non-drying of the finish material.

Preparation
 (i) *Do not* abrade as the resultant scratches may increase the degree of bleeding.
(ii) Ensure that the surface is free of dust and grease.

Primer   Aluminium sealers or proprietary anti-bleed sealers. Shellac or 'Stop Tar' knotting is satisfactory only for small areas.

Resinous and oily woods

Softwoods and hardwoods containing an excessive amount of natural resin or oil which may slow the drying of paints applied to them and cause discolouration by bleeding (see 3.6). Common types are: teak, pitch pine, Oregon pine, Douglas fir.

Preparation
  (i) Ensure that the surface is dry.
 (ii) Punch nails, etc., below surface.
(iii) Wash the surface free of excessive resin or oil with cellulose thinner or white spirit.
(iv) Dry-abrade to a clean, smooth surface.

Primer   Aluminium wood primer will prevent:
 (i) discolouration by the resin (see 3.6).
(ii) natural oils affecting the hardening of the coating.

Note   Aluminium paint will prevent bleeding. It will not prevent exudation of resin.

3.5 NATURAL WOOD FINISHES

Because of their attractive grain and natural colour many hardwoods and some softwoods are not required to be painted. Their natural beauty can be preserved by protecting them with transparent finishes.

3.5.1 PREPARATION OF NATURAL WOOD SURFACES

Previously treated sound surfaces
(a) Previously varnished, lacquered or coated with a catalysed finish, in good condition and to be recoated with similar material.
(b) Previously lacquered or coated with a catalysed finish, in good condition and to be coated with an oil or alkyd varnish.

Treatment Wash and wet-abrade (see 3.3.2).

Previously treated unsound surfaces
(a) Having old coatings which are cracked, discoloured, flaking or otherwise defective.
(b) Previously coated with a spirit varnish or wax, to be recoated with similar or other material.
(c) Previously varnished to be recoated with cellulose lacquer or a catalysed material.

Treatment Complete removal with spirit paint-remover (see 3.3.1). Wax polish can be removed by scouring with steel wool and white spirit.

New or stripped surfaces
Dry-abrading by hand or orbital sander is necessary to remove possible plane marks, or to smooth raised or rough grain. It is essential when hand-abrading to rub with the grain to avoid scratching.

Stained surfaces may require either scraping, using a universal scraper, followed by abrading, or, if stains are very deep, bleaching, for which two types of material are available:
(a) Proprietary bleaching solutions which require to be left for from 6 to 24 hours, depending on type, timber or stain.
(b) Bleach site-mixed by dissolving 50 g oxalic acid crystals in $\frac{1}{2}$ litre water (fast-acting but poisonous and must be used with care).

Surfaces must be thoroughly dry-abraded after bleaching and rinsing.

Making good
Stopping under a spirit-carried finish or spirit stain-tinted hard-stopper, plastic wood or tinted stopper (see 3.2). If water-carried stain or finish is to be used, only tinted stopper is suitable.

Filling, when necessary, should be carried out with an oil-based rubbing filler (see 3.2) or vinyl-based grain filler.

Staining
Staining may be necessary if
  (i) the grain of the timber is required to be exaggerated or darkened;
 (ii) the natural colour of the timber is not desirable;
(iii) the timber requires to be colour-matched to existing joinery.

Five principal stains are available:
(a) Chemical stain – the application of warm solutions of Epsom salts, washing soda, lime or ammonia. The effect varies with each type of wood and these stains are difficult to control to produce a specific effect.
(b) Water stains. ⎫
(c) Spirit stains. ⎬ see 2.5.3
(d) Oil stains. ⎭
(e) Tinted catalysed finishes – which combine stain and finish in one operation. Tend to obscure grain.
(f) MVP coatings (see 2.5.2).

3.5.2 FINISHES

Preservative
Material having toxic properties that resist wood rot, wood-boring insects and general decay.

(a) Creosote, dark brown and cannot be subsequently coated with paint or varnish (see 2.5.1).
(b) Non-bleeding types, considerably more expensive than creosote but available in a large range of colours (see 2.5.1).
(c) MVP stain/varnish (see 2.5.2)

Shellac varnish
The most common types are button polish and French polish. They provide a high sheen surface which is brittle and offers limited resistance to moisture (see 2.5.8).

Oil, alkyd or one-pack polyurethane varnish
Slow-drying coating having excellent protective qualities for both interior and exterior woodwork. Tends to yellow on ageing (see 2.5.5 and 2.5.6). A minimum of four coats required on external surfaces.

Oil sealer
Prepared principally for floors.

Nitro-cellulose lacquer
Sprayed material which dries rapidly and produces very clear, water-resistant films for interior and exterior woodwork (see 2.3.8).

Clear finish
Usually two-pack material based on polyurethane, epoxy, melamine or urea. Produces very hard finish which is resistant to abrasion, water and many chemicals. Most types are produced in qualities suitable for floors (see 2.4.7 and 2.5.6).

Emulsion glaze
Very clear, quick-drying finish for interior woodwork. Tends to raise the grain (see 2.5.4).

Wax polish
Beeswax polish, sometimes containing silicones to improve water-resisting properties. Produces a matt finish which does not resist abrasion or solvents and is readily marked by water.

3.6 SURFACE AND PAINT FILM DEFECTS (TABLES 3.11 AND 3.12)

TABLE 3.11 **Surface defects**

Defect	Cause	Treatment
EFFLORESCENCE (Plate 7)	See Chemically active surfaces (3.4.5)	
LICHEN, MOSS AND ALGAE Growths of small, thickly growing, non-flowering plantlife.	Plantlife thriving on damp surfaces	(i) Sterilise with proprietary toxic solution (ii) Remove by scraping (iii) Sterilise surface with proprietary toxic solution (N.B. avoid splashes on skin, and consumption by children and animals)
MOULDS AND FUNGUS (Plate 8) Multi-coloured spots or patches on infected areas	A variety of airborne spores which multiply and feed on the organic matter in paints, papers, pastes and surface deposits. Promoted by dampness and poor ventilation. Common in bakeries, breweries and damp buildings and under vinyl wall coverings hung with unprotected paste. Often found behind curtains and in cupboards where ventilation is poor	(i) Trace and remedy cause of dampness (ii) Sterilise area with proprietary fungicidal solution or 3:1 mixture of water and household bleach (iii) Remove mould and fungus by scraping and washing (iv) Sterilise area with proprietary fungicidal solution or 3:1 mixture of water and household bleach (v) Observe for 7 days for any reapearance of the mould (vi) Coat with paints containing a fungicide (N.B. avoid splashes of the fungicide on the skin and accidental consumption by children and animals)
DRY ROT (Plate 9) A fungus which destroys timber Early stages: wood excessively wet, spongy and covered with silvery white branching strands (hyphae), which in extreme dampness may form a fluffy white mass. Characteristic musty, damp smell Advanced stages: wood totally dry and brittle, with no strength, cracked in a cube pattern across the grain, and darkened.	Airborne spores of dry rot germinate on wood under the following conditions: (a) Wood with more than 20% moisture content (see 3.4.3) (b) Humid atmosphere (c) Poor ventilation *Life cycle:* (1) The spores send out hyphae in search of nourishing moisture and cellulose which is sucked from the wood until it is dry and brittle	(i) Trace and remedy the cause of dampness Increase ventilation (ii) Cut out all infected timber one metre past the termination of the hyphae and *burn immediately* (iii) Treat adjacent timber with wood preservative Treat all replacement timber with preservative (iv) Wall plaster through which the hyphae may have travelled must be removed

TABLE 3.11 **Surface defects** – *continued*

Defect	Cause	Treatment
At a very advanced stage the fungus develops a fruiting body with a brick red centre	(ii) The hyphae carry moisture to dry, sound wood to enable fresh spores to germinate and so spread the growth (iii) A plate fungus forms with a red centre which contains millions of fresh spores which further contaminate damp wood	and the brickwork treated with dry rot killer and thoroughly scorched with a blowlamp before replastering. This is a job for the specialist – rarely carried out by painter.
WET ROT (Plate 10) A fungus which destroys timber. Wood is darkened and cracked with the grain. It becomes spongy and excessively wet beneath a seemingly sound veneer	Airborne spores of wet rot germinate on very wet wood (see 3.4.3) *Moisture penetration:* commonly found in window sills and the bottom of doors, windows and gates	(i) Trace and remedy cause of dampness (ii) If attack is not too severe the removal of the source of dampness is often sufficient to kill the rot (iii) If attack is severe, cut out infected timber and treat as for dry rot
PATTERN-STAINING (Fig. 3.24) Seen as a light and dark pattern in rectangular form, corresponding with the beams, rafters and joists of ceilings or mortar joints on building block walls. Also seen on walls above radiators.	Air in a room always contains floating dust. The warm air rises by convection current carrying the dust and depositing it on the ceiling. Most dust is deposited on the cooler areas	There is little a decorator can do to cure this problem but any form of heavy covering will help to insulate the surface and reduce the defect, e.g. lining and anaglypta or polystyrene tiles. Much better results can be achieved by fitting up to 50 mm of fibre glass or other insulative material between rafters
	Steel: dark areas will be immediately under steel beams as these are colder than the space between	Insulate the ceiling surface with a heavy covering such as 'Superglypta' or expanded polystyrene
	Wood: dark areas will be between timber rafters because the timber presents a warmer surface than the space between	As above, or fill the space between the rafters with an insulating material such as 50 mm thick glass fibre, expanded polystyrene or vermiculite, to even up the surface temperature

Fig. 3.24 *Pattern-staining*

TABLE 3.12 **Paint film defects**

Defect	Cause	Prevention
BLEACHING A 'whitening' or a complete change of colour of the paint film due to chemical action	Action of acid fumes upon pigments such as ultramarine blue	Correct choice of paint and colours to suit the environment
	Action of alkali upon pigments such as prussian blue and brunswick green	Use of alkali-resisting primer on alkaline surfaces
BLEEDING Discolouration of the finishing paint by some constituent of the surface being dissolved by the paint medium	Painting surfaces coated or contaminated with bitumen, oils or nicotine	Use aluminium paint as a sealer or sizing and lining
	Painting surfaces or paints containing lake pigments or dyestuffs	Use aluminium sealer
	Painting knotty or resinous woods	Use shellac or aluminium paint as a sealer

TABLE 3.12   **Paint film defects** – *continued*

Defect	Cause	Prevention
BLISTERING (Plate 11) Eruption of a paint or varnish film forming bubbles on the surface	Trapped moisture in wood. Made worse by heat, particularly dark colours	Use moisture meter to make sure timber is dry
	Resin exuding from knots due to heat or sunlight	Replace resinous knots with sound timber
	Trapped moisture in plaster and rendering	Use permeable coatings or wait until surface is dry
	Using non-heat-resistant paints adjacent to sources of extreme heat	Use heat-resistant paint. Light colours have more resistance to blistering by heat than dark colours
CHALKING Formation of a powdery deposit on the surface of dry paint film, the powder being unbound pigment	Paint deficient in binder because of overthinning	Thin paints only when specified
	Painting insufficiently-sealed absorbent surfaces	Seal absorbent surfaces and touch-up filling
	Highly pigmented spirit-borne paints on exterior exposure	Correct choice of paint
	Use of interior paints on exposed surfaces	Correct choice of paint
CISSING (Plate 12) Failure of paint or varnish coating to form a continuous film on the surface. The film rolls back in globules, leaving small round bare patches	Painting greasy surfaces	Thorough washing and rinsing of the surface
	Painting very smooth, shiny surfaces	Abrade to remove shine
	Finishing over an oil undercoat	Do not adulterate undercoats Lightly wet-abrade
CRACKING OR CRAZING (Plate 13) Splitting of the surface due to the top coat of paint or varnish being unable to expand to the same degree as the previous coatings	Applying hard drying coatings over soft, elastic coatings	Choice of correct paint system
	Recoating before undercoat has dried	Allow specified drying time before recoating
	Adding excessive driers	Add only minimum amount of driers
	Paste and size over new paintwork	Remove all paste and size from paintwork immediately after paperhanging

TABLE 3.12 **Paint film defects** – *continued*

Defect	Cause	Prevention
FLAKING OR PEELING (Plate 14) Breakdown of adhesion resulting in paint or varnish lifting away from the surface in flakes	Painting damp surfaces, especially wood	Use moisture meter to make sure timber is dry
	Painting powdery, friable surfaces, e.g. distemper	Correct preparation and sealing
	Painting in humid conditions which promote condensation	Do not paint in foggy, wet or frosty conditions
	Expansion and shrinkage of painted surfaces	Select paints to suit nature of surface
	Formation of rust under paints	Ensure thorough preparation and correct primer
	Lack of adhesion on smooth, shiny surfaces	Wash and wet-abrade surface
	Efflorescence forming under the paint film (see 3.4.5)	Remove all efflorescence by dry-brushing before painting
LOSS OF GLOSS AND SINKING Failure of a paint or varnish film to maintain its potential or original sheen or shine	Painting insufficiently sealed absorbent surfaces	Seal absorbent surfaces and touch-up filling
	Use of wrong paint system	Use correct primers and undercoats
	Wet paint film affected by damp weather or condensation	Plan the work so that paint will dry in correct conditions

3.7 PREPARATION OF SURFACES FOR GRAINING, MARBLING AND SURFACE TEXTURING

All surfaces to be surface textured must be:

(i) Hard. Allow at least two or three days for the ground coat to dry.

(ii) Free from surface identation. The surface must be thoroughly filled, otherwise scumble will collect in the holes and show-up as darker areas.

(iii) Free from grit. All coatings must be strained and the dried ground coat abraded with 400 grade abrasive paper.

(iv) Free from brushmarks.

(v) Ground coat of slight sheen. An eggshell finish is ideal.

(vi) Full ground coat. No misses or scuffs in the ground coat which will be accentuated by the scumble.

(vii) Ground coat of correct colour. This should usually be a tone lighter than the lightest part of the desired finish.

(viii) Free from grease. Wash with detergent and rinse. This is particularly essential when using water colour. If cissing still occurs, a slurry of fullers earth, whiting, or fine pumice and water, rubbed-in over the surface will ensure good wetting.

3.8 PREPARATION OF SURFACES FOR GILDING AND BRONZING

Metallic leaves and powders can be applied to any dry and hard surfaces:

Ferrous metal
 (i) Degrease, remove all rust and coat with a rust inhibitive primer (see page 146)
 (ii) Allow the primer to dry at least 48 hours before sizing.

Non-ferrous metal
 (i) Degrease and wipe clean (see page 147).
 (ii) Priming is not essential, but if the surface is exposed, one coat of etch primer is recommended.

Timber
 (i) Thoroughly dry abrade to remove all fibres and to smooth rough grain.
 (ii) Because gold leaf will not bridge over open pores, it is necessary to fill the grain.
 (iii) A minimum of one coat of oil based primer and one coat of oil/resin undercoat is necessary to ensure a non-absorbent surface suitable for sizing.

Plaster
 (i) Must be thoroughly dry and free from dust and residue.
 (ii) A minimum of one coat of plaster sealer and one of oil/resin undercoat to ensure a non-absorbent surface suitable for sizing.

Painted surfaces
 (i) Must be clean by washing with detergent and rinsing.
 (ii) Free from texture by lightly abrading with 400 grade waterproof abrasive paper.
 (iii) Thoroughly hard, without surface defects, and free from moisture.
 (iv) Surfaces which have recently been gloss painted or varnished, and are to be sign-written or painted with decorative motifs should be treated also by one of the following processes:

Pouncing Dabbing over the entire surface to be gilded with a small muslin bag (pounce bag) containing french chalk. This absorbs any slight tackiness on the surface that may adhere the gold outside the sized areas. Surplus chalk should be brushed off before sizing.

Glair coating Glair is the white of an egg mixed with 0.5 litres of warm water. It is applied to the whole area to be gilded or bronzed, without misses, and allowed to thoroughly dry. The gold size is applied to the glaired surface. Immediately the gilding or bronzing process is completed, the surface must be washed with warm water to remove all traces of glair and any surplus metallic leaf or powder.

PART FOUR
Surface Coverings

4.1 Preparation of surfaces for surface coverings 161

4.2 Adhesives 163

4.2.1 Starch paste 163
4.2.2 Cellulose paste 163
4.2.3 Starch ethers or starch based 163
4.2.4 PVA and acrylic adhesives 164

4.3 Preparatory papers 165

4.3.1 White lining paper 165
4.3.2 Brown lining paper 165
4.3.3 Reinforced, cotton-or linen-backed lining paper 165
4.3.4 Waterproof or pitch paper 165
4.3.5 Metal foil 166
4.3.6 Expanded polystyrene 166

4.4 Flat papers 167

4.4.1 Grounds 167
4.4.2 Pulps 168

4.5 Embossed papers 169

4.5.1 Dry embossed paper 169
4.5.2 Wet embossed paper 169

4.6 Relief papers 171

4.6.1 Ingrains: woodchip, oatmeal 171
4.6.2 Plastic paint print 171
4.6.3 Relief or blown vinyls 171

4.7 Washable papers 172

4.7.1 Ordinary washable paper 172
4.7.2 Vinyl paper 172
4.7.3 Paper coated after hanging 172
4.7.4 Pre-pasted (ready-pasted) paper 173
4.7.5 Photo murals 173

4.8 Decorative laminates 174

4.8.1 Flock papers 174
4.8.2 Wool strand and weftless materials 174
4.8.3 Paper backed hessian 174
4.8.4 Paper backed felt 175
4.8.5 Grasscloth 175
4.8.6 Cloth backed vinyl 175
4.8.7. Silk 176
4.8.8 Glass fibre 176
4.8.9 Lincrusta 176
4.8.10 Metallic foils 177
4.8.11 Cork 177

4.9 Defects of surface coverings 178

4.10 Glossary of terms 182

4.11 International performance symbols 185

4.1 PREPARATION OF SURFACES FOR SURFACE COVERINGS (Tables 4.1 and 4.2)

Stripping old paper

All previous covering and all traces of adhesive and size must be removed to reveal a clean, sound surface by either dry-scraping, soaking and scraping, or steam stripping.

Sizing

Increased adhesion of coverings to a surface can be obtained by the application of either a glue size solution, when using a starch paste, or a thin coat of adhesive, when using cellulose or PVA adhesive. Sizing also reduces the absorbency of porous surfaces, so preventing 'snatch'.

Nail or screw fixings

Timber building boards, plasterboards and hardboards are often fixed with nails or screws. To prevent these fixings corroding they should be touched-up with metal primer before the surface is primed and sized.

Oil priming paints

Some surfaces require an oil primer before sizing and hanging coverings. The reasons are:
(i) to reduce surface porosity;
(ii) to bind down any existing emulsion or oil-bound distemper coatings;
(iii) to make subsequent stripping easier.

Degreasing

Previously painted, non-porous and new concrete surfaces require degreasing with detergent

TABLE 4.1 Preparation of bare surfaces

Surface	Ensure material is thoroughly dry	Wash with detergent and rinse	Dry scrape	Dry abrade	Touch up nail heads	Fill joints, nail holes and cracks	Apply linen tape over joints and fill edge	Apply appropriate primer/sealer	Size
Concrete	X	X	X	X		X			X
Expanded polystyrene	X					X			
Gypsum plaster	X		X	X		X			X
Hardboard	X				X	X	X	X	X
Insulating board	X				X	X	X	X	X
Lime plaster	X		X	X		X			X
Plasterboard (dry lining)	X			X	X	X	X	X	
Wood building board	X			X	X	X	X	X	X

TABLE 4.2 **Preparation of previously treated surfaces**

Surface	Wash with detergent and rinse	Soak and wash off surface coating	Strip wallpaper and wash wall	Dry scrape and abrade	Wet abrade	Fill joints, cracks and imperfections	Apply penetrating sealer	Size	Apply lining paper	Apply reinforced lining paper
Emulsion painted				X		X	X			
Painted tongued-and-grooved boarding	X				X	X		X		X
Oil painted	X				X	X		X	X	
Papered plaster			X	X		X		X		
Size bound distempered		X		X		X		X		

before hanging wallpaper. New concrete surfaces are greasy because of the use of release oil when casting from moulds or shuttering.

Etching or abrading

Previously painted and other non-porous surfaces should be roughened by abrading to provide extra mechanical adhesion for the wallpaper adhesive. Common practice is to abrade with soda block or Grade 150 waterproof silicon carbide paper ('Wetordry') during the washing process.

Jointing tape

A 50 or 75 mm linen reinforced or paper tape is stuck over the joints of timber building boards, plasterboard and hardboards before hanging wallpaper. This prevents joints from opening which might cause the wallpaper to split.

Lining paper

Lining paper adheres firmly to the surface and provides a ground of even porosity and absorption for the wallpaper. Surfaces should be cross-lined before hanging heavy wallpapers or hand-printed papers, or hanging to previously oil-paintedsurfaces. Heavy lining should be used on badly cracked surfaces, and reinforced lining on surfaces likely to move, e.g. tongued-and-grooved boarding.

Special surface characteristics

Dry lining is a timber framing clad with plasterboard with its decorative side exposed. Joints should be caulked and taped (see 3.2). The surface should be sealed with an oil primer to prevent damage to the paper surface when stripping subsequent coverings. *Damp walls* Hanging wallpaper to damp walls may cause bleaching of the wallpaper colours, disintegration of the paper or mould growth. The treatment is:
(i) trace and cure source of dampness;
(ii) if cure is not possible and dampness is not severe, apply either pitch paper, metal foils or proprietary damp-proofing solution.

Efflorescence is a white, fluffy crystalline salt deposit on the face of alkaline materials (see 3.4.5) which must be rubbed off.

Mould growth is seen as small multi-coloured spots or patches on infected wall or surface covering (see 3.6).

The treatment is:
(i) remove infected wallpaper and burn;
(ii) apply a fungicidal solution (see 3.6).
(iii) use fungicidal adhesive to hang surface coverings, particularly vinyls.

4.2 ADHESIVES

The range of adhesives for surface coverings can be broadly divided into four groups.

4.2.1 STARCH PASTE

Swellable starches from maize, corn, potato, wheat, tapioca or manioc. Available in powder or ready-to-use form.
Properties
High solid – low water content.
Opaque.
Advantages
Long open time.
Good bonding properties to substrate and material.
Disadvantages
High marking property (easily stains face of paper).
Encourages and supports mould growth.
Many contain preservative to reduce risk of early putrification.

Cold-water paste powder
Easily prepared and suitable for all types of wall and ceiling papers except vinyls. Should be mixed fresh daily, as stale paste is more inclined to stain.

Hot-water paste powder
Inconvenient where boiling or very hot water is not readily available. Must be allowed to cool before use as hot paste will oversoak the paper. Very glutinous. Suitable for all types of wall and ceiling papers, hessian, corks, grass-cloths, but not vinyls. Must be used fresh.

Prepared or tub paste
Requires no preparation other than dilution to suit weight of paper. A very smooth paste. May be acidic, affecting metallic prints; even if washing soda is added, some discolouration may occur. Similar in use to hot-water paste.

Dextrine
A very stiff yellow maize glue. Has great strength. Available in powder or ready-prepared form. Suitable for anaglypta panels, lincrusta mouldings, or for strengthening other pastes.

4.2.2 CELLULOSE PASTE

Combination of various cellulose ethers obtained from cotton or wood pulp. Thin, quick-setting paste, suitable for lightweight papers only.
Properties
Low solid – high water content.
Transparent.
Advantages
Good affinity with paper, as it is of similar origin.
Low marking quality (unlikely to stain face of paper).
Does not encourage or support mould growth.
Disadvantages
Higher wetting properties than starch pastes.
Lower adhesive properties than starch pastes.
Insufficient *immediate* adhesion to anchor vinyls to surface, therefore not recommended for these papers.

4.2.3 STARCH ETHERS OR STARCH BASED

Chemically produced starch ethers combined with alkali-treated cellulose.
Properties
Medium solid content, more than cellulose and less than starch.
Whitish paste which is partially transparent.
Usually available in fine flake form.
Advantages
Mixes easily to a smooth paste.
Easy to apply.
Similar setting time to starch pastes.
Better adhesive qualities than cellulose.

4.3.5 METAL FOIL

Paper backed aluminium foil.
Size   10.05 m long × 530 mm wide.
Use   To protect decoration from isolated damp patches and indelible stains activated by moisture.
Hanging
  (i) Applied with special adhesive supplied by manufacturers of the foil.
  (ii) Overlap joints by at least 3 mm.
  (iii) Avoid contact with electrical switches and socket outlets.
Note   For small isolated damp patches or stains, lead foil or aluminium foil can be used and stuck to the surface with thick oil paint.
Decorating   Metallic foils produce difficult surfaces upon which to adhere paper, but optimum adhesion can be obtained if the surface is cross lined first.

4.3.6 EXPANDED POLYSTYRENE

Foam polystyrene resin cut into thin sheets. Being 98 per cent air it has exceptional heat-insulation properties, but is highly flammable. Can be painted with emulsion paint or covered with paper if a decorative finish is required.
(a) Rolls between 0.5 and 2 mm thickness.
(b) Tiles, sheets or panels available in a large range of thicknesses from 3 to 25 mm and densities from 'coarse grain' to 'very fine grain'. Tiles are available with embossed or printed designs or textures, and with square or bevelled edges.
Size
Rolls   600 mm wide × 9 m long.
Panels   Large range of sizes from approximately 900 to 1600 × 1200 mm.
Tiles   150 mm square, 225 mm square, 300 mm square, 300 × 450 mm and 300 × 600 mm.
Use
 (i) To reduce thermal conductivity of walls and ceilings, cutting down possibility of condensation and pattern-staining.
 (ii) To prevent water penetration of damp surfaces.
Hanging
Rolls   (i) Applied with stout fungicide paste; (ii) must be butted; (iii) smooth with brush: a roller will damage the surface.
Tiles and panels   (i) Applied with stout fungicide paste or special PVA adhesive; (ii) apply adhesive over entire surface to reduce fire-spread.
Disadvantages
  (i) Being very soft, it is easily and permanently damaged by pressure.
  (ii) Dissolved by most strong solvents, although resistant to alcohols.
  (iii) Not to be used on surfaces whose temperature exceeds 80°C.

4.4 FLAT PAPERS

4.4.1 GROUNDS

Paper completely coated with a casein-bound paint before being printed with a repeating pattern or texture. Non-washable surface.

chintz jaspe

Fig. 4.2 *Machine prints*

Fig. 4.1 *Machine printing*

Machine prints (Fig. 4.1)
Repeating pattern, overall surface textures or stripes printed by rollers and containing up to 22 colours. The method by which most papers are produced.

Some grounds have a particular character and may be known by other titles. The most common are:
Jaspe A fine irregular vertical texture similar to that of scumble texture dragging (Fig. 4.2).
Chintz Pattern based on natural forms treated in a realistic manner to imitate the cotton furnishing fabric of the same name (Fig. 4.2).
Satin Ground colours polished or glazed to produce a sheen before being printed.
Moire Satin or satinette ground, finely textured to produce a watered silk effect, most commonly produced as vinyl.
Satinette Fine mica flakes incorporated in ground colour to produce a sheen.
Metallic Bronze or aluminium powder inclu-

ded in ground colour or print to produce metallic sheen.

Hand prints
Good quality ground coated papers printed by a hand process. Considerably more expensive than machine prints. Some have a selvedge (see Glossary of terms, 4.10). There are two principal types:
Block prints (*Fig. 4.3*) Printed by a method using hand-operated carved wooden blocks.

Fig. 4.3 *Block printing*

Fig. 4.4   *Screen printing*

Screen prints (*Fig. 4.4*)   Each colour of a design is squeezed out through a stencil design which is mounted on nylon material or metal mesh stretched tight over a printing frame. This method is also carried out mechanically and may be classified as a machine print.

Size   BS 1248 : 1990 states that wallpaper shall be 10.05 m long × 530 mm wide. A plus or minus tolerance of 5 per cent is permitted on the width, which has resulted in many standard papers being made available in 520 mm widths.

Use   Because the size of the pattern repeat is restricted by machine (maximum about 500 mm) machine prints constitute the bulk of wall and ceiling papers used for domestic purposes.

Hand prints are available in longer repeats, therefore more adaptable for use on very large areas and where cost does not restrict their use.

Hanging
  (i) Applied with any stout starch or starch ether paste. Cellulose pastes can be used with lightweight papers.
 (ii) Joints should be butted.
(iii) May be smoothed with brush or felt roller.
(iv) If paste penetrates through paper or accidentally gets on to the face of some dark prints it can result in a patchy sheen or colour change.

4.4.2 PULPS

Self-coloured papers upon which the pattern or texture is printed direct *without* the surface being completely coated first. Virtually obsolete.

Although paper quality and design quality may vary they constitute the lowest price range of a manufacturer's collection.

Size   10.05 m long × 520 or 530 mm wide.

Use   Rarely used for other than domestic purposes.

Hanging
  (i) Applied with cellulose, starch or starch ether pastes.
 (ii) Will quickly absorb paste and become difficult to handle if oversoaked.
(iii) Joints should be butted.
(iv) May be smoothed with brush or felt roller.

4.5 EMBOSSED PAPERS

Papers which have been pressed into a relief texture so that the back of the paper is hollowed.

4.5.1 DRY EMBOSSED PAPER

Printed paper passed between embossing rollers: either a steel roller with the pattern or texture in relief presses the paper into a soft roller, or male and female rollers are used (Fig. 4.5).

Fig. 4.5 *Embossing machinery*

(a) *Embossed* The most common and cheapest range, either with all-over random relief texture or with the pattern in relief. Depth of relief may decrease considerably after hanging.
(b) *Duplex* Two papers bonded together before being embossed, to produce a more pronounced relief. Usually the pattern only is embossed.
(c) *Low relief anaglypta-type* A duplex paper, embossed while the adhesive bonding the two papers is still wet, which helps to retain maximum relief after hanging. The top paper is a craft type for strength and whiteness. These papers differ from those described above as they are not coloured or

printed with a pattern. Generally available either in textures such as pebbledash, plaster swirl, broken glass or in geometrical patterns. Available in white only and can be painted after hanging.

Size 10.05 m long × 520 or 530 mm wide.

Use Similar use to grounds. Texture tends to mask slight irregularities on the surface which is a common use for anaglypta.

Hanging
 (i) Applied with stout starch or starch ether paste.
 (ii) The relief should not be filled with paste.
 (iii) Duplex papers require soaking until supple. Oversoaking may cause delamination.
 (iv) Use of seam roller will flatten the texture and spoil the result.

4.5.2 WET EMBOSSED PAPER

A much heavier quality paper than anaglypta, prepared from cotton linters, rosin size, china clay and alum. While in a wet state it is moulded between a steel roller and its counter made of gutta-percha. The wet-moulding ensures that its relief is retained after hanging. Usually available in white only, and suitable for subsequent painting in oil-or water-based paint.

(a) *Supaglypta-type* Available in rolls and a range of textures similar to anaglypta.
(b) *High relief panels* Very heavy moulded paper available only in panels, usually in imitation of brick or stonework. Relief may be up to 25 mm in depth. The hanging technique is entirely different from that of other embossed papers.

Size

Supaglypta 10.05 m long × 520 mm wide.

Panels Vary according to design: approximately 0.5–0.75 m².

Use

Supaglypta Often used in domestic or industrial situations to mask irregularities on badly marked surfaces. Subsequent painting provides durable and washable finish.

Panels Used for decorative effect domestically, in restaurants, hotels and film and television sets.

Hanging

Supaglypta

 (i) Applied with stout starch or starch ether paste, or prepared paste.

(ii) Requires to be soaked until supple before hanging.

(iii) Joints should be butted.

(iv) Smooth with paperhanging brush.

 (v) Seam roller will flatten relief and spoil the result.

High relief panels

 (i) Soaked with water before adhesive is applied.

(ii) Fixed with dextrine, which is knifed on to areas which come in contact with the surface.

(iii) Held to surface with steel pins until adhesive has set.

(iv) Usually designed to overlap or interlock.

4.6 RELIEF PAPERS

Papers having a relief texture on the face, but with a smooth backing. There are three principal types: ingrains, plastic paint prints and relief or blown vinyls.

4.6.1 INGRAINS: WOODCHIP, OATMEAL

Paper which has small particles of wood or cork sandwiched between a heavy backing paper and a thin surface paper.

Size   10.05 m long by 520 or 530 mm wide.

Use   Grades having very fine particles may be machine-or hand-printed and have uses similar to grounds.

The coarser grades (woodchip) are supplied in natural colour and subsequently painted with oil-or water-based paints. Their texture tends to mask irregular surfaces.

Hanging
  (i) Applied with any stout starch or starch ether paste.
 (ii) Oversoaking may cause delamination, or make paper difficult to handle without damage.
(iii) Joints should be butted.
 (iv) May be smoothed with felt roller or brush.

4.6.2 PLASTIC PAINT PRINT

A ground printed with a colour thickened with plaster or similar filler. Virtually obsolete.

Size   10.05 m long by 520 or 530 mm wide.

Use   Similar to grounds.

Hanging
  (i) Applied with any stout starch or starch ether paste.
 (ii) Over-soaking may cause print to soften.
(iii) Excessive brushing may cause print to break or leave surface.
 (iv) Pattern may crack if sharply folded or carelessly handled.
  (v) Joints should be butted.

4.6.3 RELIEF OR BLOWN VINYLS

A paper-backed material having an expanded coating of PVC which is textured in a variety of random or repeating patterns. Usually available in a neutral colour suitable for painting with alkyd or emulsion paint after hanging.

Size   10.05 m long by 530 mm wide.

Use   Domestic or industrial surfaces to provide interesting textures with good durability.

Hanging   As vinyl papers.

4.7 WASHABLE PAPERS

Papers which have a surface that can be cleaned without damaging the pattern.

4.7.1 ORDINARY WASHABLE PAPER

A ground or embossed paper which has been machine-coated with a clear glaze based on polyvinyl acetate (PVA). It will withstand regular wiping with a damp sponge, but not usually saturation washing or scrubbing with brush or abrasive.

Size 10.05 m long × 520 or 530 mm wide.
Use For surfaces which may be subject to handling or rubbing, or where condensation is not excessive, e.g. ventilated kitchens or bathrooms. Principally used for domestic purposes.
Hanging
 (i) Applied with any stout starch or starch ether paste which contains a fungicide.
 (ii) Joints should be butted.
 (iii) May be smoothed by brush or felt roller.
Note Some types of washable paper are manufactured as loosely-bonded duplex papers to facilitate removal prior to redecoration. The top coated paper can be pulled away, leaving the plain backing which can be stripped off by soaking and scraping or left as a lining paper for subsequent coverings.

4.7.2 VINYL PAPER

Smooth or textured thin film of polyvinyl chloride (PVC) bonded to a paper backing and printed with PVC inks heat-fused on to the surface. Provides a completely washable surface which is also resistant to abrasives and hard wear. Has anti-static properties, therefore tends to repel dust.

Size 10.05 m long × 520 or 530 mm wide.
Use For surfaces which are subject to condensation, considerable handling and rubbing, or splashing and staining. The wide range of patterns available makes them suitable for both domestic and industrial use.
Hanging
 (i) Because of its impervious surface, pastes dry very slowly; unless they contain a powerful fungicide, mould growth will result. It also has a tendency to pull away from the surface when drying, and the paste must be particularly strong to resist this pull. Special adhesives are available which incorporate both these properties.
 (ii) Joints should be butted.
 (iii) May be smoothed with brush, rubber roller or plastic squeegee.
 (iv) Special adhesive is necessary where paper has to be lapped, as paste will not adhere to vinyl.
Note Removal of vinyl paper prior to redecoration is carried out by pulling away the vinyl film from the paper backing. The paper can be left as a lining for subsequent coverings, or removed by soaking and scraping.

4.7.3 PAPER COATED AFTER HANGING

Most pulps, grounds and embosses which are not produced with a washable finish can be coated after hanging has been completed and the paste has dried.

The coating that is used is based on PVA (see 2.5.4) and is brushed, rolled or sprayed on to the paper. Some papers are printed with poorly bound colour which may lift or smudge if coated. Before sealing, the paper should be tested for its suitability for this treatment.

Usually the sheen produced is greater than that of a washable paper, although the degree of gloss can be controlled by thinning the glaze with water.

Fig. 4.6 *Pre-pasted paper*

4.7.4 PRE-PASTED (READY-PASTED) PAPER (Fig. 4.6)

Vinyl or washable paper which has been coated on the back with a water-activated fungicidal paste. More expensive than similar non-pasted papers but this may be offset by the saving of pasting time.

Size   10.05 m long × 520 mm wide.

Use   Similar to vinyl and washable papers.

Hanging
  (i) After being cut into lengths, the paper is soaked in trough of water.
 (ii) Joints should be butted.
(iii) Smooth with a sponge.

4.7.5 PHOTO MURALS

Very large photographs often consisting of several panels to obtain the full picture. Each panel has an overlap to allow the joints to be trimmed and butted.

Some are normal bromide photographic paper, others are on special resin coated paper.

Size   Single or multi-panelled units up to 1.270 m wide.

Use   Feature panels in a range of commercial and domestic offices and buildings.

Hanging
   (i) Applied with tub paste or PVA adhesive.
  (ii) Plan layout and joints before hanging multi-panelled murals.
 (iii) Trim surplus leaving a gap of about 10–15mm.
 (iv) Important that light source shines into the joint.
  (v) Trim with knife and straight-edge.
 (vi) Smooth with screen printer's squeegee working from the centre outwards.
 (vii) Bromide papers need soaking in water for 5–10 minutes until pliable.
(viii) Always keep surface wet otherwise the squeegee will damage the surface.

4.8 DECORATIVE LAMINATES

Paper, fabric and fibre glass backings onto which is laminated a range of natural and synthetic decorative materials.

Note Cross lining is recommended as the preparation for all paper backed decorative laminates.

4.8.1 FLOCK PAPERS

Originally made to look like brocade or cut velvet fabrics. Many are still made with traditional patterns.

Hand made The pattern is block printed or screen printed onto a paper ground using a slow setting adhesive. Short fibres of wool, silk or synthetic fibre about 2 mm long (flock) are blown on, sticking to the adhesive. This gives a raised pattern known as the pile.

Machine made The majority of flocks are now machine printed onto a variety of backings including vinyl and metallics. By using synthetic adhesives and fibres, washable flock papers can be produced.

Double flocks Higher relief, sometimes in two tones of colour. Produced by applying one tone of flock onto parts of another. Patterns can also be made by using different coloured fibres or different coloured adhesives.

Storage The pile can be crushed if stored laying down giving a shady finish when hung. Always stand on end to store.

Size 10.05 m long × 520 mm wide.

Use Decorative appearance used in domestic buildings, restaurants and older type manors and large houses.

Hanging
  (i) Applied with tub paste or PVA adhesive.
 (ii) Trim by undercutting with knife and straight-edge, ensuring that the straight-edge is placed on the selvedge of the paper so as not to damage the flock.
(iii) Do not crease after pasting.
(iv) Joints should be butted.

 (v) Smooth with felt roller, rolling downwards.
(vi) Paste on the face can cause matted fibres which are impossible to rectify.
(vii) After hanging, brush the pile into one direction with a soft brush to give an even coloured appearance.

4.8.2 WOOL STRAND AND WEFTLESS MATERIALS

Separate strands or threads of wool or other material laminated to a paper backing. There is no weft, and they can be of an overall plain colour or produced in a variety of coloured stripes.

Size
  (i) 10.05 m long × 525 mm wide.
 (ii) 50 m long × 1.0 m wide.

Use General decorative finish mainly in domestic buildings and offices.

Hanging
  (i) Applied with stiff tub paste or PVA reinforced paste.
 (ii) Very little soaking necessary.
(iii) Trim with knife and straight-edge.
(iv) Joints should be butted.
 (v) Smooth with felt roller.
(vi) Cleanliness is important, paste removed from the face can damage the fibres.

4.8.3 PAPER BACKED HESSIAN

Made from jute which is either dyed and woven into a cloth, or woven into a cloth and then dyed. The hessian cloth is laminated onto the paper. Some types have a very even texture. Others have random joints in the warp and weft known as slubs to give more texture and a pronounced feature to the finish.

Size
  (i) 10.05 m long × 520 mm wide.
 (ii) 50 m long × 910 mm wide.

Use General decorative finish on offices, public and domestic buildings.

Hanging

(i) Applied with stiff tub paste or PVA reinforced paste.

(ii) Stiff paste and short soaking time prevents undue stretching and reduces the risk of delamination.

(iii) Trim with sharp knife and straight-edge to achieve a clean first time cut. A blunt knife or poor trimming will cause frayed edges.

(iv) Reverse alternate lengths to reduce the effects of edge to edge shading.

(v) Smooth with felt roller.

(vi) Joints should be butted.

(vii) Cleanliness is important, paste can mark the face of the cloth.

4.8.4 PAPER BACKED FELT

Made by laminating felt to a paper backing. Felt is made by compressing a layer of short wool fibres lightly packed and interlocked in random directions like a blanket.

Size

(i) 10.05 m long × 520 mm wide.

(ii) 50 m long × 910 mm wide.

Use General decorative finish on offices, public and domestic buildings.

Hanging

(i) Applied with stiff tub paste or PVA reinforced paste.

(ii) Stiff paste and short soaking time prevents undue stretching and reduces the risk of delamination.

(iii) Trim with sharp knife and straight-edge to achieve a clean first time cut. A blunt knife or poor trimming will cause frayed edges.

(iv) Reverse alternate lengths to reduce the effects of edge to edge shading.

(v) Smooth with felt roller.

(vi) Joints should be butted.

(vii) Cleanliness is important, paste can mark the face of the cloth.

4.8.5 GRASSCLOTH

Originally made in Japan but is now produced in other far eastern countries.

Made from dried grasses, reeds, split bark, and split canes which are sometimes dyed. The grasses are woven with silk or cotton thread to form a random design which is adhered to a paper backing. Some types have the background showing through as part of the design, others have the whole surface dyed.

Size 7.135 m long × 910 mm wide. They can vary in size.

Use Expensive and exclusive wall coverings, very fragile and difficult to clean. Mainly used in domestic buildings and offices.

Hanging

These papers vary a great deal and manufacturer's instructions should be carefully adhered to.

(i) Applied with a stiff tub paste.

(ii) Do not soak. If the cloth becomes wet it will delaminate from the paper.

(iii) Trim with knife and straight-edge or a Ridgley track and wheel trimmer.

(iv) Matching is impossible and joints will show. Plan the area to achieve a balanced effect.

(v) Joints should be butted.

(vi) Smooth with a felt roller.

(vii) Soft grasscloths will turn into and around angles like other wallhangings. Stiff grasscloths must be cut to fit exactly into internal angles, and partially cut through from the face on external angles.

4.8.6 CLOTH BACKED VINYL

Plain or textured wallcovering made by coating or laminating polyvinyl chloride (PVC) onto paper or scrim cloth. They can be self coloured, textured to imitate fabrics, and have designs printed with special inks. They generally provide a tough waterproof and easily cleaned surface.

Size Usually 30 m long × 1.3 m wide, but can be purchased by the metre run.

Use For decorative effects and surfaces which have to take hard wear, such as corridors, staircases, offices, and public buildings.

Light colours are discoloured by fumes from gas fires and tobacco smoke.

Hanging

These materials are waterproof and the adhesive must dry out through the wall.

(i) Applied with PVA adhesive. All pastes

must contain a fungicide, or be of a type which does not support mould growth.

(ii) The paste can be applied either to the back of the material, or to the wall depending on manufacturer's instructions.

(iii) Joints should be overlapped and cut through, or the material can be trimmed with knife and straight-edge and butt jointed.

(iv) Smooth with a plastic or flexible metal spatula.

(v) Where overlaps cannot be avoided, use special overlap adhesive to adhere edges.

(vi) Any paste on the surface is best left to dry and then removed by stiff dry brushing.

4.8.7 SILK

Silk yarn dyed and woven with an extremely fine warp. The weft is usually thicker and joints in the yarn are rolled together producing a pronounced marking in the weave called a slub. The silk is then laminated onto a paper backing. The natural shimmer of the silk creates a unique sparkling effect.

Size

(i) 7.30 m long × 910 mm wide.

(ii) 6.85 m long × 760 mm wide.

Use Expensive and exclusive wall covering, very fragile and difficult to clean. Used mainly in domestic building and offices where it will receive little wear and tear.

Hanging

(i) Surface preparation is critical as any surface defects will be highlighted by the sheen of the silk.

(ii) Applied with tub paste used sparingly.

(iii) Trim with knife and straight-edge.

(iv) Matching is not possible and the joints will show. Plan the area to achieve a balanced effect.

(v) Do not soak, if the material becomes wet it will delaminate from the paper backing.

(vi) Smooth with a felt roller.

(vii) Joints should be butted.

(viii) Trim around obstacles with scissors not knives.

(ix) Cleanliness is vital as paste on the surface cannot be sponged off. Thin paste, oversoaking, and water on the face will cause delamination.

4.8.8 GLASS FIBRE

Made from very fine fibres of glass either by weaving into coarse, medium, or fine open woven material, or by pressing into a fine random mesh called chopped strand, or fine surfacing tissue.

Use A fire resistant material used (i) to reinforce cracked and imperfect surfaces, and applied to shuttered concrete to provide a flat even texture; (ii) as a finishing material over coloured grounds to produce a two coloured relief texture.

Size 50 m long × 1.0 m wide and purchased by the metre run.

Hanging

(i) PVA adhesive applied to the wall a little short of the width of the material.

(ii) The joints are lapped and cut dry with knife and straight-edge. The material is then peeled back and the wall pasted beneath the joint.

(iii) Smooth with felt roller, plastic or metal spatula.

(iv) The edges can be trimmed dry before hanging, and the walls pasted and material hung in the usual way.

4.8.9 LINCRUSTA

Stout backing paper coated with a putty like compound which is rolled with textured rollers to produce a range of ready finished relief patterns and textures such as wood and hessian.

Size

(i) 10.05 m long × 525 mm wide

(ii) Also available in panels.

Use Hard wearing material used for high traffic areas and for rich decorative effects in domestic, commercial and public buildings.

Hanging

(i) Applied with manufacturer's special adhesive.

(ii) Soak the paper backing with warm water for 20–30 minutes to make the material pliable.

(iii) Trim by undercutting with knife and straight-edge.

(iv) Internal angles must be accurately cut and butt jointed. External angles can only be turned on well rounded surfaces, otherwise cut and butt joint.

(v) Smooth with cloth and rubber roller.

(vi) Remove excess adhesive immediately with damp sponge.

4.8.10 METALLIC FOILS

Made by either laminating a paper backing with tarnished leaves of various metals, or by spraying metallic or coloured polyester resin onto a paper backing. They can be overprinted with a range of designs.

Size   7.30 m long × 910 mm wide. Other sizes also available.

Use   Rich decorative effects in exhibition display and feature areas of domestic and office decoration.

Hanging
  (i) Applied with PVA adhesive.
  (ii) Advisable with some polyester coated papers to paste the wall twice to obtain a better surface.
 (iii) Trim with knife and straight-edge.
 (iv) Joints should be butted.
  (v) Fold carefully without creasing.
 (vi) Smooth with felt or rubber roller.
 (vii) Remove any paste from the surface immediately and polish with a soft cloth.
(viii) The mirror like finish of these papers highlights every surface defect, therefore the surface preparation must be perfect to obtain an acceptable effect.

4.8.11 CORK

Thin veneer of natural cork laminated to coloured or metallic paper backing. Holes occur naturally in the thin cork veneer and the coloured or metallic backing shows through to add to the decorative effect.

Size
 (i) 7.30 m long × 910 mm wide.
 (ii) 10.05 m long × 530 mm wide.

Use   Quite hard wearing and used in domestic and office buildings.

Hanging
  (i) Applied with tub paste or PVA reinforced paste.
  (ii) Trim with knife and straight-edge or Ridgley track and wheel trimmer.
 (iii) Dampening the edges with water before pasting will sometimes prevent excessive curling of the edges which is common to these papers.
 (iv) Joints should be butted.
  (v) Smooth with felt roller.

Note   Cork is also available in tile and sheet form. Refer to manufacturers' hanging instructions.

4.9 DEFECTS OF SURFACE COVERINGS (TABLES 4.3–4.7)

TABLE 4.3 **Surface coverings: defects due to surface or room conditions**

Cause	Visible defect
Alkaline surfaces which are still active and not sealed may bleach or discolour printing inks	Staining Discolouration
Cold damp room causes paste to dry slowly	Staining Blistering
Condensation causes slow drying of the paste, encourages mould growth or discolours the paper or printing inks	Blistering Mould growth Staining
Damp walls have similar effect to condensation	Blistering Mould growth Staining
Efflorescence prevents adhesion, darkens the paper or chemically affects the pigments in the printing inks	Peeling Edge springing Staining
Excessive porosity tends to absorb adhesive, leaving covering loosely adhering and difficult to slip into alignment	Edge springing Peeling Poor matching
Exposure to excessive sunlight may cause certain pigments to fade or discolour	Discolouration
Non-porous surface (gloss paint, tiles, plastic) which has not been lined causes the paste to dry slowly, resulting in the paper fibres stretching	Blistering Staining
Poorly adhering lining paper is further weakened by both the wetting action of the paste and the weight of the covering	Blistering Peeling
Traces of old size and paste left unremoved before hanging vinyl	Mould growth
Uneven surface causes paper to be moulded to the undulations, resulting in stretching	Creasing Poor matching Joints gapping
Unsealed steel pins on timber or wall board corrode and stain the paper	Rust spotting Staining
Unsealed water-soluble stains dissolved by the paste	Staining

TABLE 4.4   **Surface coverings: defects due to paste and pasting**

Cause	Visible defect
Careless pasting may result in paste getting on to face of paper	Staining Discolouration Sheen patches Polishing (particularly at seams)
Hot paste penetrates paper quickly	Staining Delamination Tearing
Incorrect paste may cause undersoaking or oversoaking, reduce adhesion, prevent slip, make paper difficult to handle, or encourage mould	Blistering Peeling Edge springing Discolouration Poor matching Mould growth
Insufficient soaking makes smoothing difficult and causes uneven tensions on paper	Blistering Edge springing
Misses cause uneven swelling	Blistering Overlapping Poor matching
Oversoaking causes excessive expansion and difficulty of handling, and may separate duplex papers	Staining Delamination Blistering Poor matching Tearing
Stale paste becomes yellower, highly acidic, and may contain bacteria. It becomes thinner as it ages	Staining Darkening of metallic print Mould growth Peeling
Thin paste has reduced adhesive properties and high wetting power	Peeling Staining Tearing Joint springing
Uneven application or pasting parts twice causes uneven swelling	Blistering Overlapping Poor matching

TABLE 4.5   **Surface coverings: defects due to hanging**

Cause	Visible defect
Blunt tools	Tearing Inaccurate finishing
Careless handling may damage the paper or transfer paste and dirt to face of paper	Staining Tearing
Excessive brushing, rolling or squeegeeing causes covering to stretch and may polish surface of matt papers	Poor matching Creasing Polishing Loss of emboss
Insufficient care in shading	Variations in colour of lengths
Insufficient smoothing	Blistering Joint gapping Creasing
Reversing alternate lengths when not recommended by manufacturer	Apparent variation in colour and texture of adjacent lengths
Use of seam roller on embossed papers	Flattening of relief
Use of seam roller too soon after hanging or without protecting paper	Paste staining Polishing

TABLE 4.6   **Surface coverings: defects arising in manufacture**

Cause	Visible defect
Inaccurate trimming	Poor matching
Poorly bonded duplex paper	Blistering Delamination
Shaded papers due to use of different tones of inks or effect of light and dirt during storage	Variations in colour of lengths Darkening or lightening of edges

TABLE 4.7 **Surface coverings: summary of common defects and their causes**

Cause	Blistering	Creasing	Darkening of metallic print	Delamination	Discolouration	Edge springing	Inaccurate angle cutting	Joint gapping	Loss of emboss	Mould growth*	Overlapping	Paste staining	Peeling	Polishing	Poor matching	Rust spotting	Shading	Sheen patches	Staining	Tearing
Alkaline surface					X														X	
Cold damp room	X																		X	
Condensation	X									X									X	
Damp surface	X									X									X	
Efflorescence						X						X							X	
Excessive porosity						X						X	X							
Excessive sunlight					X															
Non-porous, unlined surface	X																		X	
Poorly adhering lining	X											X								
Uneven surface		X						X							X					
Unsealed steel pins/screws																X			X	
Unsealed water-soluble stain																			X	
Careless pasting					X						X		X					X	X	
Hot paste			X																X	X
Incorrect paste	X					X	X	X					X		X					
Insufficient soaking	X					X														
Misses in pasting	X								X				X							
Oversoaking	X			X									X						X	
Stale paste		X							X			X							X	
Thin paste								X				X							X	X
Uneven pasting	X								X				X							
Blunt tools							X													X
Careless handling		X																	X	X
Careless seam rolling											X									
Excessive brushing	X	X							X		X			X	X					
Insufficient smoothing	X	X						X												
Reversing alternate lengths																	X			
Rolling embossed paper									X											
Inaccurate trimming															X					
Poorly bonded duplex				X																

*Mould growth is not always apparent on the face of surface coverings although it may be far advanced beneath the material. Particularly on papers, it may easily be detected by staining. Advanced moulds may be detected by smell, particularly in humid conditions.

4.10 GLOSSARY OF TERMS

Batch or shade numbers (Fig. 4.7) An identification number printed on the back of the material at the end of the roll or on a label inserted beneath the transparent cover. Indicates the code number of the colour mixed to

Fig. 4.7 *Batch numbering*

print the paper; only coverings having the same number are likely to be exactly the same colour.

Butt jointing The most common method of hanging surface coverings: the edges touch without a gap or overlap.

Centring Method of setting out a wall, ceiling or panel, particularly where a large pattern is being used, to avoid an unbalanced effect. The two alternative methods are shown in Fig. 4.8.

Fig. 4.8 *Centring*

Fig. 4.9 *Concertina folding*

Concertina folding (Fig. 4.9) Method of folding long lengths of pasted covering to be applied to ceilings or horizontally to walls. Sometimes used when hanging very long lengths vertically.

Fig. 4.10 *Use of a crutch*

Crosslining The hanging of lining paper horizontally: the normal method when subsequent covering is to be hung vertically.

Crutch (Fig. 4.10) A support used when hanging ceiling papers to prevent the folded pasted paper bending or creasing. May be a roll of paper, cardboard roll or a straight-edge.

Cutting top and bottom (Fig. 4.11) Method

Fig. 4.11 *Top and bottom cutting*

of marking and cutting ends of lengths after application to surface to ensure accurate fitting into ceiling and skirting angles.

Double lining The hanging of two layers of lining paper, one vertically and the other horizontally. A method used to mask very irregular surfaces.

Fig. 4.12 *Drop pattern*

Drop pattern (Fig. 4.12) A pattern which does not repeat on a horizontal line from edge to edge of the material. Requires special care when cutting lengths, to ensure that matching is accurate. Also offset match.

Fig. 4.13 *End-to-centre folding*

End-to-centre folding (Fig. 4.13) Method of pasting coverings which are to be hung vertically. Ensures that no pasted areas are exposed, makes for easy carrying and delays set of paste.

Lapping The overlapping of edges of consecutive lengths of surface covering by up to 20 mm. Mainly used when applying pitch papers or metal foils, to minimise the risk of moisture penetrating through the joint.

Length Part of a roll cut to fit the surface to be covered. Usually calculated as the exact distance between ceiling angle and skirting, or both wall angles when covering ceilings, plus an allowance of about 100 mm to ensure accurate marking and cutting at angles (see Cutting top and bottom).

Matching The method of aligning the pattern of one length of surface covering to the preceding one.

Plumb first length Because not all wall angles or door and window openings are perfectly upright it is necessary to hang the first length on each wall to a line that has been set out exactly vertical (see Plumbing).

Plumbing The use of a plumb bob and line or large spirit level to produce an exact vertical line.

Random pattern (Fig. 4.14) A design that has no apparent repeat therefore does not require matching. Textures, stripes and jaspes come within this group. Some of these are called 'semi-plain'. Also called free match.

Fig. 4.14 *Random pattern*

Release agent A coating containing waxes, applied to the prepared surface to make the subsequent removal of paper easy.

Reversing alternate lengths In some non-patterned or semi-plain coverings, slight variations in colour may be masked by hanging every other length in the opposite direction. This technique should be used only when recommended by the manufacturer.

Roll or piece The smallest unit by which surface coverings can be purchased. Size may vary according to country of origin (see Standard size).

Selvedge (Fig. 4.15) A narrow strip (approx 12 mm) on each edge of a roll of surface covering, permitting accurate printing. Usually removed by the manufacturer, but

Fig. 4.15 *Selvedge*

may be left on some hand prints and lincrusta (see Trimming).

Shading Checking that each roll of the same pattern is exactly the same colour. Batch numbers must be the same, and the rolls compared visually.

Sizing The application of a thin coat of glue size or paste to a porous surface to ensure even absorbency, before hanging surface coverings (see 4.1).

Slip The degree to which a covering can move once it is applied to a surface. This property is crucial to accurate matching.

Smoothing The method of removing creases and air bubbles after the covering has been applied and of ensuring good contact with the surface.

Snatch The holding of the applied covering by a too-absorbent surface, allowing no movement. The opposite to slip.

Soaking length The period of time that a length of heavy surface covering requires to be left after pasting and folding before it becomes sufficiently soft and supple to be hung.

Standard size English papers: 10.05 m long by 530 mm wide. A plus or minus tolerance of 5 per cent is permitted on the width, which has resulted in many standard papers being made available in 520 mm width. Foreign papers vary in both length and width; sizes must be checked before estimating.

Star cut (Fig. 4.16) The normal method of cutting coverings to fit around obstructions on the surface which cannot be removed.

Fig. 4.16 *Star cut*

Straight or set pattern (Fig. 4.17) A design that repeats in a horizontal line, the match being at the same level on each edge.

Fig. 4.17 *Straight or set pattern*

Striking a line Method of producing on a ceiling or sloping surface, using a chalked string, a line against which to hang the first length.

Trimming Removal of the selvedge by the use of either a knife and straight-edge or a hand trimmer before pasting, or with a track and wheel cutter after pasting and folding.

Fig. 4.18 *Turning corners*

Turning corners (Fig. 4.18) The lapping of wall coverings round corners. Few wall angles are perfectly true, and if the lap is more than about 15 mm, shrinkage on drying may cause blisters.

Working from the light An alternative method of working when centring is not used. If lengths are hung starting by the chief source of natural light, any slight unavoidable overlapping will not cast a shadow and will be less apparent.

4.11 INTERNATIONAL PERFORMANCE SYMBOLS

A range of internationally recognised symbols which often appear in wallpaper pattern books and on product labels.

TABLE 4.8 **International performance symbols**

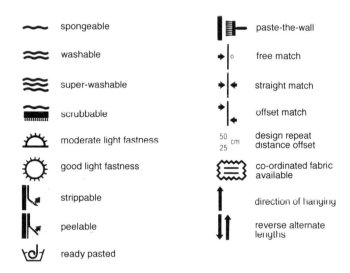

spongeable	paste-the-wall
washable	free match
super-washable	straight match
scrubbable	offset match
moderate light fastness	design repeat distance offset
good light fastness	co-ordinated fabric available
strippable	direction of hanging
peelable	reverse alternate lengths
ready pasted	

Reproduced by permission of the Wallcovering Manufacturers' Association.

PART FIVE

Scaffolding

5.1 Ladders, trestles, stepladders 189

5.1.1 Ladders 189
5.1.2 Ladder brackets and other
     attachments 190
5.1.3 Roof ladder 190
5.1.4 Trestles 191
5.1.5 Stepladders 192
5.1.6 Splitheads 192
5.1.7 Hop-ups 193

5.2 Boards, staging, working platforms 194

5.2.1 Scaffold boards 194
5.2.2 Lightweight staging 194

5.3 Scaffolding 196

5.3.1 Tubular components 196
5.3.2 System or unit scaffold 198
5.3.3 Mobile scaffold 199
5.3.4 Independent scaffold 201
5.3.5 Putlog scaffold 201

5.3.6 Birdcage scaffold 202
5.3.7 Slung scaffold 203
5.3.8 Truss-out scaffold 203

5.4 Suspended scaffolds 205

5.4.1 Cradle 205
5.4.2 Bosun's chair 207

5.5 Ropes, knots 208

5.5.1 Types of rope 208
5.5.2 Knots 208

5.6 Safety equipment 210

5.6.1 Safety belts, harnesses 210
5.6.2 Safety nets and sheets 211

5.7 Statutory obligations 213

Note In Part Five Scaffolding, all references to regulations about the construction and safety of scaffolding are taken from the Construction (Health, Safety and Welfare) Regulations 1996 which replace the Construction (Working Places) Regulations 1996, the Construction (Health and Welfare) Regulations 1996, and the Construction (General Provisions) Regulations 1961.

5.1 LADDERS, TRESTLES, STEPLADDERS

5.1.1 LADDERS

Ladders are used for work which is easily accessible and of a light nature. They are suitable for work which is too high for trestles or stepladders, or where large patent or tubular scaffolds cannot be used. Pole ladders are usually used for access on tubular scaffolds.

Extension ladder

Constructed of wood, aluminium alloy or a combination of both, i.e. wood stiles with metal rungs.

Timber stiles should be straight-grained, free from knots and have a wire rope set into the back for strength. They should be protected by varnish rather than paint, which might hide defects. Rungs are either rectangular or round and made from hardwood such as hickory, oak or ash. Metal rods are fixed beneath the rungs at regular intervals to prevent the stiles parting (Fig. 5.1a).

The stiles of metal ladders should be constructed of rectangular box section of heat-treated aluminium alloy. The rungs are usually ribbed for a better grip (Fig. 5.1 c).

Extension ladders can be of double or triple sections, hand-push or rope-operated, using various patent methods of rung locking to prevent collapse (see Fig. 5.3).

Pole ladder (Fig. 5.1 b)

Single-length ladder with stiles made from one straight tree trunk sawn in half lengthways. This ensures that both stiles have even spring. All timber should have the bark stripped from it. Construction, shape and reinforcement of rungs is similar to that of extension ladders.

Storage (Fig. 5.2)
(i) Laid flat.
(ii) Supported in at least three places along its length.
(iii) Stored under cover but not in a dry store.

Safety precautions (Fig. 5.3)
(i) *Never* use ladders with missing or loose rungs or with makeshift repairs to rungs.

a *Timber ladder* b *Pole ladder*
c *Aluminium ladder*

Fig. 5.1

Fig. 5.2 *Ladder support*

Fig. 5.3 *Securing the ladder*

(ii) *Never* repair broken stiles with wooden splints nailed or lashed to them.

(iii) *Never* stand ladders on uneven, soft or loose ground.

(iv) *Never* reach out too far: it is easy to overbalance.

(v) *Never* try to gain extra height by resting the ladder on an unstable surface, or supporting it on its rungs.

(vi) *Never* climb a ladder beyond 1 m from the top.

(vii) *Always* secure the ladder
    (a) at the top, to prevent it from slipping; or, if this is not practical,
    (b) to anchorages at or near ground level; if neither of these is possible,
    (c) by a person standing on the bottom rung.

(viii) *Never* use ladders with frayed or damaged ropes.

(ix) *Always* use ladders at the correct angle.

(x) *Only* use ladders according to their Duty Rating, see 5.2.3.

5.1.2 LADDER BRACKETS AND OTHER ATTACHMENTS

Ladder brackets (Fig. 5.4)
Metal supports which are hooked over two rungs of a ladder, used in pairs to support a working platform. Usually made from 6×25 mm steel sections with adjustable support and rung hooks.

Fig. 5.4 *Ladder scaffold*

Use Suitable for use where a single ladder does not give adequate access and the work does not involve the use of heavy materials. They support lightweight staging of 3 m maximum span.

Safety precautions

(i) Use only brackets which are supported on two rungs.

(ii) Both ladders must be lashed or wedged to prevent slipping.

(iii) Guard rails are essential for platforms over 2 m above ground.

Maintenance
Grease or oil all moving parts.
Keep free from corrosion.
Replace pivot pin when worn.

Safety feet (Fig. 5.5)
Rubber or plastic padded metal feet which can be fitted to the bottom of ladder stiles to prevent them slipping. They are hinged to allow the ladder to be used at the correct various angle.

Fig. 5.5 *Safety feet*

Ladder stay (Fig. 5.6)
A metal bracket attached to the top of a ladder, used to hold the ladder away from the building for easier working and access to projecting or overhanging features such as gutters, eaves and soffits.

Fig. 5.6 *Ladder stay*

5.1.3 ROOF LADDER (Fig. 5.7)

Made from wood or aluminium, at least 380 mm wide with cross bars 32 mm thick fixed at 380 mm intervals. Wheels are fitted to slide up the roof and hooks to anchor the ladder over the ridge.

Fig. 5.7   *Roof ladder*

Use
(a) To gain access on to and over sloping and pitched roofs (e.g. having a pitch of more than 10°), to prevent slipping.
(b) To avoid fracturing fragile materials such as asbestos, slate or glass.

Safety precautions (Fig. 5.8)
  (i) When working on fragile roofs, two ladders should be used side by side.
 (ii) Access ladders to a roof should where possible be fixed to a window brace or to a roof rafter.
(iii) When roof slope is over 30° or is dangerously slippery, one of the following structures must be erected at the eaves.
     (a) a catch barrier or platform
     (b) a working platform with guard rail.
(iv) A safety belt and rope fastened to a safe anchorage should be worn.

5.1.4 TRESTLES (Fig. 5.9)

Made from rectangular sectioned softwood or aluminium alloy section and wide enough to take two scaffold boards or one lightweight staging. The cross members are spaced approximately 500 mm apart and staggered on each side to give a platform rise of approximately 250 mm. At least two tie rods should be fitted to both sides of timber trestles for extra strength. The timber should be protected with varnish rather than paint so that defects are not hidden from view.

They are available in a range of sizes between 2 and 4.5 m high.

Storage   Closed, with check blocks located in the opposite stile, and standing under cover.

Safety precautions
  (i) *Never* use trestles when only partly open.

Fig. 5.8   *Safety measures for roofwork*

Fig. 5.9   *Folding trestles*

(ii) *Never* use a trestle in a closed position as a ladder.

(iii) Guard rails and toe boards are required for platforms over 2 m high.

(iv) *Never* use folding trestles for more than one tier.

(v) At least $\frac{1}{3}$ of any folding trestle should be above the working platform.

(vi) *Never* use trestles with missing or loose cross members.

(vii) *Never* use trestles with loose or slack hinges.

(viii) *Always* dismantle scaffolding before moving.

(ix) Only use trestles according to their correct Duty Rating, see 5.2.3.

5.1.5 STEPLADDERS (Fig. 5.10)

Stepladders are constructed of either timber or aluminium alloy.

Timber stepladders are made from softwood. Treads should be housed into stiles at 225 mm centres and reinforced with steel tie rods at regular intervals. Rustless hinges are bolted and screwed for extra strength. Ropes are fitted to control the opening of steps when in use. The timber should be protected by varnish rather than paint, which might hide defects.

Aluminium stepladders are lighter, and fitted with metal bars to control the opening of the steps. Rubber feet are fitted to prevent slipping.

Available in sizes from 5 to 10 treads high.

Storage Closed, tied with their ropes and standing under cover.

Safety precautions

(i) *Never* use steps when only partly open.

(ii) *Never* use steps in a closed position as a ladder.

(iii) *Never* use steps with a working platform over 2 m high.

(iv) *Never* use a stepladder unless the operative's knees are below the top rung.

(v) *Never* use steps with frayed or uneven length ropes.

(vi) *Never* use steps with missing or loose treads.

(vii) *Never* use steps with loose or slack hinges.

(viii) *Never* work from the top of a stepladder unless it has been constructed as a platform with a secure hand hold.

treads housed
glued, nailed

Fig. 5.10   *Swing-backed steps*

(ix) *Always* dismantle scaffold before moving.

(x) *Always* select steps by their Duty Rating, see 5.2.3.

5.1.6 SPLITHEADS (Fig. 5.11)

Splitheads are made of metal. The height of the supporting heads can be adjusted either by a pin and hole method, or by the use of screw jacks.

Use They are used to support a close boarded working platform particularly suitable for access to a ceiling. They can be adjusted to support a platform from 770 mm to 2.6 m high.

A minimum of four splitheads are necessary to form a working platform. For large areas eight splitheads may be necessary. Timber bearers are placed on edge spanning between the splitheads, and scaffold boards or staging are supported by the bearers.

Fig. 5.11   *Splithead scaffold*

Safety precautions
  (i) Always place the working platform as close to the wall as possible to avoid user falling off the edge.
 (ii) All moving parts must work freely.
(iii) Pins must be sound and fit securely into the holcs.
(iv) Heads must not be bent.
 (v) Tripods must be firmly supported and free from rocking.
(vi) Bearers must be wedged to prevent rocking.

5.1.7 HOP-UPS
(Figs 5.12, 5.13 and 5.14)

Hop-ups can be hand built of wood or purchased ready made and are obtainable as three main types:
(a) The hand-built variety consists of a simple step up to a flat working platform measuring not less than 500 mm × 400 mm (see Fig. 5.12).

Fig. 5.12   *Hand-built hop-up*

The ready-made variety is available in two forms:
(b) Ready made, of aluminium similar to a two tread stepladder (see Fig. 5.13).

Fig. 5.13   *Ready-made hop-up*

(c) Ready made, consisting of two short ladder-type sections made of aluminium alloy which support a short lightweight staging made of aluminium and wood. They are adjustable for three working heights of between 420 mm and 580 mm (see Fig. 5.14).

Fig. 5.14   *Ready-made hop-up – adjustable*

Use A light and versatile means of access to the top section of walls not exceeding approximately 2.5 m high, or used in conjunction with scaffold boards or lightweight staging for access to ceilings of similar height.
Safety precautions
  (i) Ensure hand-built varieties are sturdy, remain steady in use and are protected only with varnish which will not hide any defects.
 (ii) Small stepladder types must be fully opened and locked before use.
(iii) Ensure the side frames of the ready-made type are firmly locked against the stops and angled outwards.

5.2 BOARDS, STAGING, WORKING PLATFORMS

5.2.1 SCAFFOLD BOARDS (Figs. 5.15 and 5.16)

Made from softwood (spruce or fir). The timber should be straight-grained, free from large knots and splits. Ends of boards should have the corners cut off and be protected from damage by binding with metal strips.

Boards must not be less than 200 mm wide if less than 50 mm thick. If thicker than 50 mm they may be 150 mm minimum width.

Available in a range of lengths, rarely exceeding 4m.

Use   To provide a working platform on ladder, trestle and tubular scaffolds.

Safety precautions

 (i) The maximum distance between supports depends on the thickness of the board: 32 mm can span 1 m, 38 mm can span 1.5 m, 50 mm can span 2.6 m.

 (ii) The minimum width of a working platform is 430 mm or two boards when used on trestles. On tubular scaffolds the working platform must not be less than 600 mm or 3 boards.

(iii) Maximum permissible gap between boards is 25 mm.

(iv) Maximum permissible overhang of boards on a working platform is four times the thickness of the board, and no less than 50 mm.

 (v) Every board must rest securely and evenly on its support.

Storage Stored under cover. Supported throughout their length on well ventilated racks and arranged according to size.

5.2.2 LIGHTWEIGHT STAGING (Fig. 5.17)

Specially constructed timber platform, of Douglas fir or hemlock, for spanning greater widths than is possible with scaffold boards. Timber should be protected with varnish rather than paint, which might hide defects.

Fig. 5.15   *Scaffold board*

not less than 3 boards 0.600 m when used as a footing only

overhang not more than 4 times thickness of board

maximum distance between transom support 1.500 m

Fig. 5.16   *Scaffold board supports*

Fig. 5.17 *Staging*

Platform supports are reinforced with steel rods, spanning between stiles. Stiles are strengthened throughout their length with wire reinforcement. The length of the stile is designed to be convex to allow for sag during use. Provides a 500 mm wide working platform up to 7.3 m long. The maximum span is 3 m. Where a span is anticipated to be or used over 3 m additional supports are necessary. The staging is designed to carry three operatives plus equipment (up to 270 kg evenly distributed load). All staging must be marked with its maximum permitted distributed loading.

Use Provides working platforms for trestle and ladder bracket scaffolds or can be used directly on the building structure (between roof trusses), so saving costly and space-consuming scaffolding.

Storage Stored upside down and supported at ends to retain the convex stile design. Lightweight staging is also available constructed of aluminium alloy stiles and cross members supporting a slatted timber platform. They do not have reinforcing wires and do not require to be stored upside down.

5.2.3 CLASSES OF PORTABLE PLANT

All ladders, steps, trestles and staging have a Duty Rating:

Class 1 *Industrial* for use on building sites.
Class 2 *Light Trades* for more general use.
Class 3 *Domestic* for household use only.

All ladders should be marked with their duty rating.

5.3 SCAFFOLDING

5.3.1 TUBULAR COMPONENTS

Scaffolding must only be erected, modified or dismantled under the supervision of an experienced and competent person.

Parts of a scaffold (Figs 5.18 and 5.19)

Standards   Upright tubes which bear the weight of the scaffold.

Ledgers   Horizontal tubes which extend along the whole length and act as supports for putlogs and transoms.

Transoms   Span across ledgers to support working platform.

Guard rail   A safety rail running not less than 0.910 m or more than 1.140 m above working platform. A mid rail leaving no gap greater than 0.470 m.

A *not more than 0.470 m if open*
B *1.000 m or over*
C *between 0.910 and 1.140 m*

Fig. 5.19   *Scaffold: safety requirements*

A *standard* B *ledger* C *transom* D *guard rail* E *toe board* F *raker*
G *traverse or cross brace* H *longitudinal or diagonal brace* J *reveal tie* K *sole plate* L *tie*

Fig. 5.18   *Parts of the scaffold*

Toe board Usually a scaffold board on edge, running along the outside of the working platform to prevent objects falling off.

Raker Inclined load-bearing tubes or buttresses to support scaffolds which cannot be otherwise tied to the building.

Traverse or cross brace Triangulates the frame to give greater rigidity to the scaffold.

Longitudinal or diagonal brace Serves the same purpose as cross brace.

Reveal tie Tubes jacked between opposite sides of an opening to provide an anchor to tie the scaffold to the building.

Sole plate Boards or sleepers forming the main bearing for base plates to give greater distribution of load.

Tie A tube used to connect the scaffold to the reveal tie or other method of tying in.

Putlog Short tube with a flat end. The flat end fits into the brick joint, the open end is coupled to a ledger. It can support a working platform (see 5.3.5).

Puncheon A vertical tube that does not rest on the ground. Used where a standard cannot be taken to the ground (see 5.3.8).

Bridle Horizontal tube fixed between putlogs to span openings in a wall and support transoms (see 5.3.5).

Tubes

Tubes may be of steel or of aluminium alloy. Steel tubes should be welded or seamless. If not galvanised they should be painted or varnished to prevent corrosion. Aluminium alloy tubes need not be painted but care should be taken to avoid contact with harmful materials, i.e. alkalis or salt water.

Steel and aluminium tubes should not be used in the same scaffold as under load they bend to different extents.

Steel and aluminium alloy fittings may be used in the same scaffold.

Scaffold fittings

Universal coupler (Fig. 5.20) A load-bearing fitting for connecting two scaffold tubes at right angles, e.g. standard to ledger.

Fig. 5.20 *Universal coupler*

Fig. 5.23 *Putlog coupler*

Fig. 5.26 *Base plate*

Fig. 5.29 *Putlog end*

Fig. 5.21 *Double coupler*

Fig. 5.24 *Sleeve coupler*

Fig. 5.27 *Adjustable base plate*

Fig. 5.30 *Toe board clip*

Fig. 5.22 *Swivel coupler*

Fig. 5.25 *Joint pin*

Fig. 5.28 *Reveal pin*

Fig. 5.31 *Final coupler*

Double coupler (Fig. 5.21)   A load-bearing fitting for connecting two scaffolding tubes at right angles.

Swivel coupler (Fig. 5.22)   A fitting used for connecting two tubes at any angle, e.g. cross brace to standard.

Putlog coupler (Fig. 5.23)   For securing (a) putlogs or transoms to ledgers to prevent movement; (b) guard rails to standards. Must not be used for load-bearing.

Sleeve coupler (Fig. 5.24)   For joining two scaffold tubes end to end.

Joint pin or expanding spigot (Fig. 5.25)   An alternative fitting for connecting two scaffold tubes end to end. Fitted internally, it expands to grip the wall of the tube.

Base plate (Fig. 5.26)   Placed under standards to distribute the load. It has an integral spigot and fixing holes for use with sole plates.

Ajustable base plate (Fig. 5.27)   For use on uneven or sloping ground.

Reveal pin (Fig. 5.28)   A fitting used for tightening a tube between two reveals to form a reveal tie.

Putlog end (Fig. 5.29)   For converting a scaffold tube into a putlog.

Toe board clip (Fig. 5.30)   For fixing a toe board securely to a scaffold standard on working platform.

Finial coupler (Fig. 5.31)   For fixing a guard rail on top of a standard.

Safety precautions   A universal or double coupler is the only type which should be used where a load has to be transferred from one tube to another via a coupler, e.g. for joining a ledger to a standard.

When not in use, fittings should be cleaned and soaked in light oil and stored in bins.

5.3.2 SYSTEM OR UNIT SCAFFOLD
(Figs. 5.32, 5.33, 5.34 and 5.35)

A patent form of metal scaffold designed to reduce the need for loose fittings, and to speed up and simplify erection. Made of either steel (usually galvanised) or aluminium alloy, the latter being lighter and so more suitable for the painter and decorator.

The various systems can be divided into three groups:

Fig. 5.32   *System scaffold. Type a*

(a) Standard scaffold tubing fixed by means of special brackets (Fig. 5.32).

(b) Frames varying in width but of standard depth, which slot into each other at right angles. These systems include guard rails, castors, outriggers and sometimes gantries to give access between units (Fig. 5.33).

(c) A simple tower unit in which all components are built in. The cross braces are hinged so that the tower can be erected by unfolding. It can be folded flat for storage after use. Lightweight platform units lock into the frame, giving extra rigidity to the scaffold. Access ladders are contained within the scaffold and adjustable legs take up changing floor levels (Fig. 5.34).

Use   Principally for erecting towers, whether used singly or in pairs to support working platforms.

Safety precautions   Safety lies only in carrying out the procedure laid down by the manufacturer at all times.

If instructions are not available the following rules apply:

Fig. 5.33 *System scaffold. Type b*

Fig. 5.34 *System scaffold. Type c*

(i) Ensure that the base is firm.
(ii) Uprights must be plumb.
(iii) All joints must be securely locked in place.
(iv) Only adequate and effective ties and bracing must be used.

Maintenance   Ensure that sections are not bent or twisted, slots and tube ends are not blocked and welds are not fractured.

5.3.3 MOBILE SCAFFOLD (Fig. 5.35)

A simple tower made from tubular scaffold components or system units and mounted on wheels. It has a small working platform and can easily be moved from place to place.

Constructed of normal steel tube and fittings or system units with the addition of locking castor wheels. The height of the tower should not exceed $3\frac{1}{2}$ times the length of the shortest base side (including outriggers if fitted), when used for internal work. When used externally it must not exceed 3 times the length of the shortest base side: for example, a tower of 2×1.5 m base must not have a working platform more than 4.5 m high if used for external work.

Bracing should extend the full height of the scaffold both vertically (zig-zag) and horizontally (plan bracing). No tower should have a base less than 1.220 m wide. A tower over 9.750 m high should be tied to the building, anchored by guy ropes or steadied by weighting the bottom.

Lifts should not exceed spacing of standards. The working platform should be close-boarded and fitted with toe boards and guard rails.

An access ladder should be provided to the working platform, long enough to provide a hand-hold 1.100 m above the platform and lashed to the scaffold to prevent movement.

Fig. 5.35 *Mobile scaffold*

Fig. 5.36 *Independent tied scaffold*

Fig. 5.37 *Putlog scaffold*

Use Internally, where access is required to ceilings, walls, columns or specific areas where detailed work is required, c.g. picking-out.

Externally, where repetitive work is carried out, e.g. to rows of windows, lamp standards or balconies.

Safety precautions

 (i) Wheels must be secured to standards and locked when in use.
 (ii) When moved, the tower should be pushed or pulled only from the base, and the working platform should be free of men and material.
(iii) Mobile towers should only be used on firm and level ground.

5.3.4 INDEPENDENT SCAFFOLD (Fig. 5.36)

The scaffold stands clear of the building and is designed to carry the full weight of all load of materials and men placed upon it. It must be tied to the building for stability and to prevent movement towards or away from the building.

When used for painting or similar work, standards may be placed up to 2.700 m apart, with 1 m between the rows. This allows for a four-board-wide platform.

Ledgers are fixed 2.600 m from the ground and then at 2 m intervals. Transoms are placed 2.400 m apart except when supporting a working platform, when the distance should be 1.200 m. The maximum height for this type of scaffold is 61 m.

Use Generally used by most building trades on external surfaces of multi-storey buildings when extensive work is to be carried out.

Safety precautions

 (i) Standards must be placed on base plates. On soft, uneven or fragile ground, sole plates must be used.
 (ii) Standards must be vertical or slightly inclined towards the building.
(iii) It is *essential* that tube joints in standards and ledgers are staggered and NOT kept in line.
(iv) Boards must be evenly supported.
 (v) Working platforms must have guard rails and toe boards.
(vi) Safe means of access must be provided throughout the scaffold.

Maintenance Like all tubular scaffolds, they must be inspected every seven days and immediately after a spell of bad weather.

5.3.5 PUTLOG SCAFFOLD (Fig. 5.37)

The putlog scaffold has only one row of standards and depends for stability on supporting

2.000 m
to
2.400 m

2.600 m max.

2.400 m max.

Fig. 5.38 *Birdcage scaffold*

itself by means of the wall of the building. It may be extended vertically as the building itself is built. Maximum height 46 m.

The standards are spaced 1.800–2.400 m apart, depending on the type of work. They are placed not less than 1.270 m and not more than 1.320 m away from the wall. The working platform must not exceed 5 boards wide. Ledgers are placed at vertical intervals of 1.370 m. Putlogs are fixed to ledgers at about 1.220 m intervals and the spade end should extend into the brickwork at least 80 mm.

Ties are very important, especially in new brickwork, and raker struts also increase stability.

Use Mainly where brick structures are being built or repaired. Sometimes called a single or bricklayer's scaffold. Used by painters on new building work.

Safety precautions As independent scaffold.
Maintenance Seven-day check, as for all types of tubular scaffold.

5.3.6 BIRDCAGE SCAFFOLD (Fig. 5.38)

An interior scaffold that provides a single working platform usually filling the room or hall in which it is being used. Maximum height 61 m.

The construction is similar to that of an independent scaffold, but rows of standards cover the whole of the work area, spaced at 2.400 m intervals. Sole plates to support base plates are essential to spread the load and protect the floor. If the floor is suspended, the sole plates should run at right angles to the floor joists.

The scaffold should be tied to the surrounding walls, beams or columns. The working platform should be close-boarded and toe boards and guard rail must be provided. Access ladder must be provided.

Use For access to ceilings and walls of large buildings such as cinemas, churches and ballrooms. Because of its high cost it is used only when the work is of long duration.

Fig. 5.39 *Slung scaffold*

Safety precautions As independent scaffold, except that standards must be upright.

Maintenance Seven-day check, as for all types of tubular scaffold.

5.3.7 SLUNG SCAFFOLD (Fig. 5.39)

A framework of tubular fittings, close-boarded and suspended from the ceiling of large buildings. It cannot be raised or lowered.

Wire ropes or chains must be used to secure the scaffold to structural members e.g. rolled steel joists (RSJs). At least six of these suspension points are required for a working platform. The ropes must be equally spaced, vertical and all taut when under load. These ropes are secured to a framework of ledgers and transoms by means of a round turn and two half hitches with the end tied (see 5.5.2). Ledgers should be at approximately 2.0 m centres and the transoms set on the ledgers at about 1.220 m centres.

The platform must be close-boarded and fitted with toe boards and guard rails. To avoid horizontal movement, guy ropes should be attached from the platform down to convenient points above ground level.

Use To provide a working platform to internal roof and ceiling areas where the floor level is not accessible or not suitable for a built scaffold, e.g. over areas of constant use such as theatres, railway platforms, work areas in factories.

Safety precautions
(i) Only wire ropes, chains or tubes must be used for suspension.
(ii) All ropes or chains must be taut.
(iii) The whole platform must be close-boarded.

5.3.8 TRUSS-OUT SCAFFOLD (Fig. 5.40)

A form of independent scaffold which is supported entirely by the building. There are two types:
(a) Horizontal cantilever tubes protrude through windows and are anchored to vertical tubes against the inside walls. The outer ends are supported by rakers bearing on the

A *raker* B *working platform* C *window opening* D *ceiling* E *floor*
F *adjustable base plate* G *puncheon*

Fig. 5.40 *Truss-out scaffold*

cills or ledges. Only suitable for very light work.

(b) Anchorage is provided inside the building by two rows of puncheons wedged between the floor and ceiling bearings and against sole plates, which should run at right angles to floor and ceiling joists. Rakers should not exceed 2.700 m in length without being braced. They should be fitted to every standard. Maximum height above the trusses 12 m.

This is a sturdy form of scaffold used for heavier work. Buildings must be in sound condition to take all the stresses and weights involved. All tubes and fittings should be new or as new condition.

Use Where access from ground level is impossible or inadvisable, e.g. small work on the upper parts of tall buildings, or over busy roads, canals or rivers.

Safety precautions As independent scaffold, but only one working platform should be used at a time.

Maintenance Seven-day check, as for all types of tubular scaffold.

5.4 SUSPENDED SCAFFOLDS

5.4.1 CRADLE (Fig. 5.41)

A working platform, suspended by ropes, which can be raised or lowered.

(a) *Standing cradle* The cradle is fixed in a predetermined position on the face of the building and can only be moved vertically.

(b) *Travelling cradle* The cradle is attached to a track fixed to a roof rig. By means of trolleys (jockeys) running in the track, the cradle may be moved both sideways and vertically; this allows complete coverage of the work area.

Use Frequently used by painters. Provide an economical working platform for extensive work on tall buildings.

Construction

Outriggers Rolled steel joists (RSJs), braced scaffold tubing, patent framework or timber poles (fir) are used as outriggers for cradles. The butt end of a fir pole should be used as the suspension point and should not extend more than 450 mm from the fulcrum point to the rope or track fixing. Outriggers must be packed with timber or similar material to avoid damaging the building.

Counterweights 25 kg weight should be secured to the tail of each outrigger to prevent over-tipping. The weights should rest on sandbags, timber or suitable padding to prevent damage to roof.

Traversing track Normally 100×75 mm alloy or steel joists in varying lengths up to 9 m, attached to the rigging by means of girder clips lashed with wire bonds. The maximum overhang of the track beyond the outrigger should not exceed 450 mm. The ends of the track must be blocked to prevent the traversing trolleys from rolling off.

Cradle or boat Constructed from steel, aluminium, timber or glass-reinforced plastic. Must

calculation of counterweight:
counterweight (A) × tail length
(C) = overhang (D) × 3 times
weight of cradle + gear + man (B)

Fig. 5.41 *Travelling cradle*

conform to BS 2830: 1994 and have the necessary close-boarded platform (with drain holes), toe boards, guard rails, mid rails and stirrups (usually made of steel). Many large buildings have traversing track or mobile davits provided on the structure for maintenance work using cradles.

Suspension system (Fig. 5.42)

Wire rope and winch   The winch may be of a type where the wire is wound around a drum, or a straight-through type where the winch climbs the wire. Steel wire is available in diameters of 6–12 mm. Wires should be kink-free, fitted with safety hooks for attachment to the suspension rig and at least 3 m longer than the height the cradle will have to travel vertically.

Rope and fall   Two single pulley blocks, one at the cradle, the other attached to the pole or to a travelling jockey on the track: the man raising the cradle needs to lift only half his own weight. The fall ropes should be of manilla three-strand, 19 mm diameter. Manilla ropes have good resistance to general wear. Synthetic ropes have

Fig. 5.42   *Suspension system*

Calculation of counterweight: counterweight (A) × tail length (C) = overhang (D) × 4 times weight of chair + gear + man (B)

Fig. 5.43   *Bosun's chair*

good resistance to chemical attack and are suitable where stone cleaning is taking place. Fall ropes must not be less than three times the height of the building plus 3 m.

Safety precautions

(i) Counterweights must be accurately calculated, using a safety factor of 3 (see Fig. 5.41):

$$\frac{(\text{length of overhang of outrigger}) \times (3 \times \text{loaded weight of cradle})}{\text{tail length of outrigger}}$$

(ii) Sandbags, water tanks or loose scrap iron must not be used for counterbalances.

(iii) All pole frameworks should be made with wire lashings.

(iv) Purpose-made stop-ends must be used on track ends: *not* wire bonds.

(v) Suspension wires must have no twists or kinks which could foul winch.

(vi) Suspension ropes must be free from fraying, deterioration and rot, and must never be spliced.

(vii) The cradle must be entered from a safe point of access, not by climbing up or down the ropes.

(viii) When not in use, cradles should be hauled up out of reach of unauthorised people.

(ix) Cradles must be inspected once a week and inspection recorded in scaffold register.

5.4.2 BOSUN'S CHAIR (Fig. 5.43)

A form of one-man suspended scaffold. The chair may be of plastic, timber or metal.

A bosun's chair is rigged with a double sheave pulley block at the top and a single pulley at the chair. Outriggers must be securely held in position or weighted at the tail. Where weights are used a safety factor of 4 is required, i.e. counterweight × tail length = overhang × 4 times loaded weight of chair.

All ropes and chains should be securely attached to the anchorage, and to the chair with a swivel connection to prevent spinning. Suspension hooks should be moused to prevent them being detached if tilted.

Use   Where it is impossible for other forms of cradle to be erected and the work is of a comparatively short duration, e.g. to fixed flag poles, stack pipes or chimneys.

Safety precautions   Bosun's chairs have only one suspension rope and one anchorage point. They are therefore more dangerous than suspended scaffolds and any mishap is serious and may be fatal. They should only be erected, inspected and used by competent and experienced people.

5.5 ROPES, KNOTS

5.5.1 TYPES OF ROPE

Three main types of rope are used by the painter: natural fibre ropes, man-made fibre ropes and wire ropes.

Natural fibre rope

Two classes of natural fibre rope are used by the painter: manilla and sisal.

Manilla   Made from fibres of certain species of tree. The fibres are long and strong. Supplied in three grades, which are identified by a blue thread as follows:

Superior quality – through centre of all three strands

Grade 1 quality – through centre of two strands

Grade 2 quality – through centre of one strand.

Sisal   Made from the leaves of a cactus. Fibres shorter than and not as strong as manilla ropes. Supplied in one grade only and identified by a red thread.

Note   If a rope has no identification it should be treated as the lowest grade.

Size   Diameter 18 and 24 mm; length 220 m.

Use

 (i) Suspended cradles and bosun's chairs.

(ii) Guy ropes and lashing.

Care and maintenance   Must be kept dry and away from all acids, gases and cleaning compounds. Wet ropes must be dried naturally as heat can cause rope to become brittle. Coil and hang on wooden pegs in well ventilated store to dry.

Man-made rope (nylon, Terylene, polyethylene, polypropylene)

Superior to natural rope in many respects, being stronger, less liable to chemical attack, completely resistant to mildew and rot, more water-resistant. Although non-flammable they may melt at high temperatures. They are 5–6 times more expensive than natural ropes.

Use   Principally for safety nets, safety belts and harnesses, and cradle suspension where chemicals are being used.

Care and maintenance   Avoid exposure to strong sunlight, heat and chemicals. If contaminated they should be hosed down with clean water and allowed to dry.

Wire rope

Wire strands twisted round a jute or hemp heart. Three sizes are used by painters:

6 mm for lashings only;

9 mm for lashings and cradles;

12 mm for slung scaffolds.

Care and maintenance   Kinked ropes or ropes with broken strands must not be used.

5.5.2 KNOTS

The knots most commonly used by the painter are the following.

Reef knot (Fig. 5.44)   For joining two ropes of similar thickness.

Clove hitch (Fig. 5.45)   For securing to scaffold tubes, outriggers and ladders.

Bowline (Fig. 5.46)   Used on safety lines to produce an eye.

Round turn and two half hitches (Fig. 5.47)   Suitable for securing ropes to tubes.

Figure of eight knot (Fig. 5.48)   A terminal knot. Useful for preventing ropes running out of pulleys and gin wheels.

Rolling hitch (Fig. 5.49)   Will withstand side-slip. Used for securing locking-rope on bosun's chair.

Sheet bend (Fig. 5.50)   For securing an end of rope to an eye or loop and for joining ropes of unequal thickness.

Tying-off (two half hitches) (Fig. 5.52)   A quick method of tying-off a cradle.

Timber hitch (Fig. 5.52)   Tightens with pressure. Used to raise and lower spars and boards.

Square lashing and hitch (Fig. 5.53)   For joining spars or boards at right angles. Used to secure ladder stiles to ledger.

Fig. 5.44   *Reef knot*

Fig. 5.49   *Rolling hitch*

Fig. 5.45   *Clove hitch*

Fig. 5.50   *Sheet bend*

Fig. 5.46   *Bowline*

Fig. 5.51   *Tying-off with half hitches*

Fig. 5.47   *Round turn and two half hitches*

Fig. 5.52   *Timber hitch*

Fig. 5.48   *Figure of eight knot*

Fig. 5.53   *Square lashing and hitch*

5.6 SAFETY EQUIPMENT

5.6.1 SAFETY BELTS, HARNESSES

Safety belt (Fig. 5.54)
A belt or harness strapped to the body and fixed to an anchored safety line.
(a) Pole belt, used for work on poles and similar work.
(b) General-purpose belt used by painters, consisting of a body belt or harness with shoulder straps. Made of man-made fibre and leather. Attached by means of a 'D' hook to a length of nylon rope which is spliced to a safety clip or hook. The length of nylon safety line should not exceed 1.800 m.
(c) Rescue harness.
Use When no other means of safety can be provided and a fall would almost certainly prove to be fatal, e.g. under high bridges, under the roofs of large railway stations, on the edge of high buildings.
Suspension points Suspension points must be strong as they take all the weight and strain. It is sometimes recommended that guidelines of steel rope be used, and the belts attached to them.

a *and* b *General purpose belts* c *Safety harness* d *Safety clip*

Fig. 5.54

Inspection and maintenance Belts and harnesses should be inspected regularly by a competent person and recorded in a register. Manmade fibres should be kept clean by washing in water. Leather belts should be cleaned and dressed at least every three months.

Mobile anchor – inertia-operated (Fig. 5.55)
A device attached to a safety line which allows free movement but locks when sudden movement or heavy weight is applied to it. Inside the device, at the top end, are three steel balls held in a cage which is part of the mass floating on steel springs. In this position the walls are free. With a downward fall the mass remains stationary and the balls are forced by a conical wedge against the cable, causing it to lock on the line.

Use When the regulation length safety line is too short or restrictive for the type of work.

Safety precautions
(i) After a fall, the cable and equipment should be checked for damage.
(ii) The anchor should be attached as close to the user as possible – never more than 600 mm.

Fig. 5.55 *Inertia-operated anchor*

Border cord: the cord surrounding the mesh which determines the size of the net.

Tie cord: secures the border cord to its supports.

Test cord: loosely threaded through the net and tested every three months to ensure that the material is not deteriorating.

5.6.2 SAFETY NETS AND SHEETS

Safety nets (Figs 5.56 and 5.57)
These are required by the Construction (Health, Safety and Welfare) Regulations 1996 when it is impractical to erect a standard working platform. They are intended to catch (a) falling personnel, (b) debris, to protect those working below.

Personnel nets normally have a mesh size of 100 mm, the mesh being either square or diamond shaped.

Debris or protection nets have a much finer mesh, average size 16 mm, and are normally used in conjunction with personnel nets to prevent injury or damage to property from objects falling from above.

Nets are designed by the manufacturer for the specific work to be carried out. Whether made of man-made or natural fibre, they must be rotproof. Terylene or nylon are the most commonly-used materials. They consist of the following parts:

Mesh cord: the cord from which the mesh is constructed.

Fig. 5.56 *Safety net*

Fig. 5.57   *Safety net in position*

Storage   Nets should be protected from damp and heat and hung on wooden pegs in a well ventilated store. If nets become contaminated with chemicals they should be washed immediately in clean water. Nets have a life of two years, after which time they should be returned to manufacturer for testing.

Safety sheets

Heavy sheeting fixed under working platforms to protect people and equipment underneath from paint spots and fine debris. Made of sound, rot-proof material such as tarpaulin, edges double-hemmed and double-stitched, all corners double thickness. All support eyelets should be brass ringed or buttonhole-stitched. Support ropes should be passed through eyelets and spliced.

Use   When working (a) over public and pedestrian areas; (b) in cinemas and factories or similar situations where people and equipment must be protected. They can be used alone or in conjunction with safety nets.

5.7 STATUTORY OBLIGATIONS

Before any type of scaffold is erected which may obstruct a public highway, an application for a licence must be made to the Highways Department of the Local Authority concerned.

The proposed site of the scaffold would be inspected by the Authority, and if appropriate, a licence issued subject to certain duties and obligations by the person applying for the licence.

The duties of the person to whom the licence is issued can be enforced by the Authority, who has powers under the *Highways Act 1980* to issue fines and penalties.

A public highway includes pavements as well as roads, and these regulations could therefore affect the use of tower scaffolds, trestles, and ladders if erected in such a way that causes any obstruction.

All Local Authorities interpret the Highways Act in a different manner, but a summary of some of the duties are as follows:

(i) To act in accordance with existing statutory regulations related to the work in progress. For example, the Construction (Health, Safety and Welfare) Regulations 1996 and/or the Workplace (Health, Safety and Welfare) Regulations 1992.

(ii) To display the licence in a conspicuous place on the scaffolding.

(iii) To ensure that the scaffolding is adequately lit during the hours of darkness.

(iv) To paint the uprights white between the ground level and 2–3 metres above.

(v) To erect and maintain safety signs warning the public of any potential hazards.

(vi) To protect the public from hazards by the use of suitable barriers.

(vii) To ensure adequate and safe passage around any obstruction.

(viii) To erect and maintain traffic signs if the safe passage includes part of the road.

PART SIX

Safety, Health and Welfare

6.1 Health and safety at work   217

6.1.1 Health and Safety at Work Act   217
6.1.2 Health and safety regulations   217

6.2 Fire precautions   223

6.2.1 Common causes of fire   223
6.2.2 Fire-fighting   225
6.2.3 Treatment for burning clothing   225

6.3 Storage of materials   226

6.4 Lead paint   229

6.4.1 Control of Lead at Work
     Regulations, 1980   229
6.4.2 Removal of paint systems
     containing lead   229

6.5 Hygiene   230

6.5.1 General principles of hygiene   230
6.5.2 Industrial dermatitis   230

6.6 Fumes, dust, noise, vibration   232

6.6.1 Fumes   232
6.6.2 Dust   233
6.6.3 Noise   233
6.6.4 Vibration   236
6.6.5 Radiation   236

6.7 Lifting   237

6.8 First aid   238

6.8.1 First-aid requirements   238
6.8.2 First-aid treatment   238

6.1 HEALTH AND SAFETY AT WORK

6.1.1 Health and Safety at Work Act

Since the 1920s, Government Statutes (laws) have existed which stated safe working conditions for building workers in general, and in some cases, painters and decorators specifically.

Because these statutes tend to take many years to pass through Parliament, and because technology in the industry is developing so rapidly, the specific regulations which have been introduced are often out of date or redundant in a short time. This situation prompted the introduction in 1974 of the *Health and Safety at Work Act* (HSWA) which made it illegal for any work site to be unsafe, and for any material or equipment to have unknown hazards. The Act does not identify any specific site conditions, equipment, or materials but directs the onus for site working conditions onto:

(i) the employer;
(ii) the employee;
(iii) the manufacturer.

Briefly, the responsibilities for each are:

The employer
 (i) Provide safe workplaces, machines, and methods of work.
 (ii) Ensure that machines and materials are used, stored and transported safely.
 (iii) Provide information, training and supervision to ensure the health and safety of the workers.
 (iv) Provide safe entrances and exits to workplaces.
 (v) Provide good, well maintained welfare facilities.
 (vi) Publish a written safety policy.
 (vii) Ensure that members of the public are not exposed to risks.

The worker
 (i) Take care of own and workmates' health and safety.
 (ii) Follow the health and safety regulations which apply to the workplace.
 (iii) Co-operate with the employer in making areas safe.

 (iv) Not misuse anything provided to make work safe.

The manufacturer
 (i) Ensure that products are without risks to health.
 (ii) Ensure that products have been fully tested.
 (iii) Provide adequate information on the safe use of the products.

HSWA is enforced by the Health and Safety Executive (HSE). They provide Health and Safety Inspectors (HSI) who have the power to enter workplaces and either advise on conditions or practices and issue Improvement Notices, or to condemn hazardous conditions and practices by issuing Prohibition Notices. HSIs can also prosecute offenders, whether they be employer, worker, or manufacturer.

The following sections identify specific situations which may prove hazardous to the painter and decorator. Although specific statute regulations are quoted, the HSWA encompasses them all, and applies equally to those which are not quoted.

6.1.2 HEALTH AND SAFETY REGULATIONS

As a result of the Health and Safety at Work Act 1974, statutory regulations are being continually updated and developed to identify the many sources of hazards in the construction industry and to give safety directions and advice to employers, employees and manufacturers/suppliers. A list of the most recent and/or important regulations for the construction worker follows:

- Control of Asbestos at Work Regulations 1987
- Construction (Design and Management) Regulations 1994
- Construction (Head Protection) Regulations 1989
- Construction (Health, Safety and Welfare) Regulations 1996
- Control of Lead at Work Regulations 1980

- Control of Substances Hazardous to Health Regulations 1994
- Electricity at Work Regulations 1989
- Health and Safety (First Aid) Regulations 1981
- Highly Flammable Liquids and Liquefied Petroleum Gases Regulations 1972
- Management of Health and Safety at Work Regulations 1992
- Manual Handling Operations Regulations 1992
- Noise at Work Regulations 1989
- Personal Protective Equipment at Work Regulations 1992
- Protection of Eyes Regulations 1974
- Provision and Use of Work Equipment Regulations 1992
- Workplace (Health, Safety and Welfare) Regulations 1992

To follow is a brief introductory guide to some of the above regulations.

Note It must be stressed that the following is only a brief introductory guide to the regulations and fuller details should be obtained from the actual regulations, approved Codes of Practice or guidance notes issued by the Health and Safety Commission and/or the Health and Safety Executive.

Many of these new regulations may overlap with existing requirements where they are still in force. Compliance with the new regulations should normally be sufficient. However, where other regulations are still in force they should be considered together with the new.

Construction (Design and Management) Regulations 1994

These regulations, known as the CDM Regulations, are intended to make clear who is responsible for health and safety matters on site, so as to ensure that health and safety is drawn into design and work methods. The CDM Regulations require the development of two key documents to ensure health and safety matters on site:

The health and safety plan

The health and safety plan provides the health and safety focus for the construction phase of a project and is developed in two stages.

1. The pre-tender health and safety plan
This is developed by the planning supervisor and should include:

(i) A general description of the work.
(ii) Details of timings within the project.
(iii) Details of risks to workers as far as possible at that stage.
(iv) Information required by potential principal contractors to demonstrate competence or adequacy of resources.
(v) Information for preparing a health and safety plan for the construction phase and information for welfare provision.

2. The construction phase health and safety plan
This is prepared by the principal contractor and should include:

(i) Arrangements for ensuring the health and safety of all who may be affected by the construction work.
(ii) Arrangements for the management of health and safety of construction work and monitoring of compliance with health and safety law.
(iii) Information about welfare arrangements.

The health and safety file

This is a record of information for the client/end user, which tells those who might be responsible for the structure in the future, the risks that have to be managed during maintenance, repair or renovation.

The planning supervisor has to ensure that the file is prepared as the project progresses and it is given to the client/end user when the project is complete. The client has to make it available to those who will work on any future design, building, maintenance or demolition of the structure.

The CDM Regulations apply to any construction work which is notifiable. This means any work which lasts for 30 days or will involve more than 500 person days of work and involves five parties:

The client
The client should be satisfied that only competent people are appointed as planning supervisor, principal contractor, designers and contractors. Duties on clients do not apply to domestic householders when they have construction work carried out.

The designer
The designer should ensure that structures are designed to avoid, or where this is not possible, to minimise risk to health and safety while they

are being built and maintained. Design includes the preparation of specifications as well as drawings.

The planning supervisor
The planning supervisor has overall responsibility for co-ordinating the health and safety aspects of the design and planning phase and for the early stages of the health and safety plan and safety file which need to be developed.

The principal contractor
The principal contractor should take account of health and safety issues when preparing tenders. They are also responsible for developing the health and safety plan and co-ordinating all activities to ensure they comply with health and safety legislation.

Contractors and the self employed
All should co-operate with the principal contractor and provide information on the health and safety risks created by their work and how they will be controlled.

Construction (Head Protection) Regulations 1989
These regulations require the wearing of suitable head protection on all building operations and construction sites, unless there is no risk of head injury other than by the person falling. Sikhs wearing turbans are exempt from the regulations.
The employer must:
(i) Assess the suitability of any head protection that is to be used.
(ii) Bring the rules to the attention of all concerned in writing.
(iii) Provide safety helmets that are suitable for the job.
(iv) Make sure that head protection is worn when instructed.
(v) Replace equipment if damaged, exposed to chemicals, or at intervals recommended by the manufacturer.
The employee must:
(i) Wear the safety helmets provided.
(ii) Take care of the equipment.
(iii) Report damage or loss of the helmet.

Construction (Health, Safety and Welfare) Regulations 1996
These regulations, known as the CHSW Regulations, are aimed at protecting the health,

safety and welfare of everyone who carries out construction work by explaining the detailed ways of working. They also aim to give protection to other people who may be affected by the work.
The CHSW Regulations replace the following regulations:
● Construction (General Provisions) Regulations 1961.
● Construction (Working Places) Regulations 1966.
● Construction (Health and Welfare) Regulations 1966.
The main duty-holders of these regulations are employers, the self employed and those who control the way in which construction work is carried out and include:
(i) Ensuring a safe place of work and safe means of access.
(ii) Preventing falls from heights.
(iii) Protecting people from falling objects and materials.
(iv) Preventing the collapse of ground around excavations.
(v) Ensuring safe traffic routes and safe operation of vehicles.
(vi) Preventing risk from fire, explosion, flooding and asphyxiation.
(vii) Providing adequate sanitary and washing facilities.
(viii) Providing site wide facilities such as fresh air, protection against adverse weather, emergency lighting, perimeter protection.
(ix) Appropriate training, inspection and reports.

Control of Substances Hazardous to Health Regulations 1994
These regulations, sometimes referred to as the COSHH Regulations, are aimed at the protection of employees and others from the effects of working with substances which are hazardous to health. COSHH does not apply to work with asbestos, lead or radioactive materials which are covered by their own specific legislation.
The employer must:
(i) Know what substances are being used.
(ii) Assess the hazards to health that the substances can cause.
(iii) Eliminate or control the hazards identified.
(iv) Give information, instructions and training to employees.
(v) Monitor the effectiveness of any controls.
(vi) Keep records.

The employee must:

(i) Make use of any control measures and equipment provided.

(ii) Comply with any arrangements the employer has put into effect.

Electricity at Work Regulations 1989

These regulations require precautions to be taken against the risk of death or personal injury from electricity and create duties for employers, self employed and employees in all aspects of electrical work.

The regulations require that no one shall do any work where technical knowledge or experience is necessary to prevent danger, or injury, unless they have the knowledge or experience necessary, or are under proper supervision as may be appropriate, having regard to the nature of the work.

Having appropriate technical knowledge means that the person must be competent. This means having:

(i) Adequate knowledge of electricity.

(ii) Good experience of electrical work.

(iii) Practical experience of the system.

(iv) Knowledge of hazards and precautions that need to be taken.

(v) Ability to recognise unsafe situations.

Information, instructions and training must be given. Written safe systems of work should form part of a 'permit to work' procedure.

Health and Safety (First Aid) Regulations 1981

These require that employers and self-employed persons provide adequate and appropriate first-aid facilities which should be based on an assessment of the need for first-aid depending on the nature of the work and not just on the number of people on site.

Employers must:

(i) Appoint a sufficient number of suitable and trained personnel to give first-aid to employees.

(ii) Appoint other persons who, in the absence of the trained first-aider, will be capable of taking charge in an emergency and looking after first-aid equipment.

(iii) Provide and maintain suitable first-aid boxes in places that can be easily reached by employees.

(iv) Display notices giving the identity of first-aiders and location of the first-aid equipment.

(v) Keep an accident report book which details all accidents.

(vi) Investigate the circumstances of all accidents.

(vii) Report notifiable events and diseases to the Health and Safety Executive.

Management of Health and Safety at Work Regulations 1992

These require that in any company, large or small, suitable assessment of the risks to health and safety of employees and others is carried out. The regulations apply to companies with as few as five or more employees. If these regulations overlap with any existing regulations to adopt the more specific regulations will normally be sufficient to comply with the Management of Health and Safety at Work Regulations. For example, the Control of Substances Hazardous to Health Regulations (COSHH) require that risk assessment takes place.

The regulations are about:

- Identifying the roles and responsibilities of all involved in the management of health and safety.
- Setting up safety standards to be achieved.
- Creation of procedures to be followed and ensuring that they are carried out.

The employer must:

(i) Assess the risks to the health and safety of employees to which they are exposed whilst at work.

(ii) Make suitable arrangements for the effective planning, organisation, control, monitoring and review of preventative and protective measures.

(iii) Provide health surveillance for risks identified in the assessment.

(iv) Appoint suitably competent persons to undertake any measures taken for health and safety.

(v) Instigate procedures for serious and imminent danger areas.

(vi) Provide information to employees.

The employee must:

(i) Use all work items correctly.

(ii) Co-operate with the employer to comply with the regulations.

(iii) Notify any shortcomings in the health and safety arrangements and take remedial action as may be needed.

Manual Handling Operations Regulations 1992

These regulations deal with safe manual handling. Where the assessment required by the Management of Health and Safety at Work Regulations (see above) indicates risks to employees from manual handling of loads, the employer must take steps to:

- Avoid manual handling as far as is reasonably practicable.
- Assess any manual handling operations that cannot be avoided.
- Reduce the risk of injury as far as is reasonably practicable.

The employer must:

(i) Identify all manual handling tasks which might involve risk of injury.

(ii) Avoid such tasks where reasonably practicable.

(iii) Where not reasonably practicable, the employer must:

    (a) Carry out a suitable and sufficient assessment of each manual handling task.

    (b) Determine and implement measures to reduce risk of injury to lowest reasonably practicable level.

(iv) Provide employees with necessary training and information concerning weight of loads, etc.

(v) Review assessments when no longer valid or changes in handling operations have happened.

The employee must:

(i) Use appropriate equipment in accordance with training and instructions.

(ii) Follow systems of work laid down by the employer.

Noise at Work Regulations 1989

These regulations set noise exposure levels and place duties on both employer and employee.

The employer must:

(i) Reduce noise to the lowest reasonably practicable level.

(ii) Have a noise assessment carried out by a competent person when exposure levels require it.

(iii) On request, provide ear protectors where noise levels cannot be reduced below 85 dB.

(iv) If noise levels cannot be reduced below 90 dB, must designate and identify, with signs, 'ear protection zones'.

(v) Ensure proper use and maintenance of equipment when it is required.

(vi) Provide information and training on the risk of damage to hearing, use of protectors and employees' responsibilities.

The employee must:

(i) Properly use any ear protectors when exposure is at, or above, 90 dB.

(ii) At all times, use any protective equipment provided and report any defects to the employer.

Personal Protective Equipment Regulations 1992

These deal with the use of personal protective equipment in the workplace which includes protective clothing such as aprons, gloves, footwear, helmets, etc., as well as protective equipment such as eye protectors, respirators and safety harnesses.

The regulations cover:

- Head protection.
- Eye protection.
- Foot protection.
- Hand and arm protection.
- Protective clothing for the body.

The employer must:

(i) Ensure suitable personal protective equipment is provided to employees.

(ii) Ensure it is appropriate to the risks involved.

(iii) Take account of the requirements and state of health of the employee.

(iv) Ensure it is capable of fitting the wearer correctly.

(v) Provide proper maintenance and repair facilities and training.

(vi) Ensure that the personal protective equipment is effective as far as is reasonably practicable.

(vii) Ensure that the personal protective equipment complies with other statutes, regulations and standards.

The employee must:

(i) Wear the equipment when provided.

(ii) Take care of the equipment.

(iii) Report damage or loss.

Provision and Use of Work Equipment Regulations 1992

These regulations, known as the PUWER Regulations, set the standard for the provision and use of work equipment. The primary objective is to provide safe equipment and to ensure its safe use so as not to give rise to risks to health and safety of the user and others. The regulations cover the full range of equipment used in construction from hammers and drills to cranes and scaffolding. They also cover equipment such as:

- Chain saws.
- Abrasive wheels.
- Angle grinders.
- Disc-cutting tools.
- Cartridge-operated tools.
- Woodworking machinery.

The employer must:

(i) Ensure that equipment is constructed or adapted for the purpose for which it is used or provided.

(ii) Consider working conditions and risks to health and safety to persons where the equipment is to be used.

(iii) Ensure that the equipment is used only for operations, and is under conditions, where it is suitable.

(iv) Provide information, instructions and training.

Workplace (Health, Safety and Welfare) Regulations 1992

These regulations apply to all workplaces, for example to shops, offices, schools, hospitals, hotels and places of entertainment. Although they do not cover construction sites, they do cover temporary work sites where much of the maintenance and refurbishment part of construction activity takes place.

The regulations are extensive and include requirements for:

- Maintenance of workplace and equipment.
- Ventilation.
- Temperature in indoor workplaces.
- Lighting.
- Cleanliness and waste material.
- Room dimensions and space.
- Workstations and seating.
- Condition of floors and traffic routes.
- Fall or falling objects.
- Windows, skylights and ventilators.
- Sanitary conveniences.
- Washing facilities.
- Drinking water.
- Accommodation for clothing.
- Facilities for changing clothes.
- Facilities for rest and eating meals.

6.2 FIRE PRECAUTIONS

Painters use many flammable materials and flame- or spark-producing apparatus. They therefore need to take precautions in the use and storage of them to prevent fire. It is essential to read manufacturers' instructions before using any material which may constitute a fire risk.

6.2.1 COMMON CAUSES OF FIRE

Highly flammable liquids (low flash point)
Usually petroleum spirit or mixtures of cellulose solution having a flash point below 32°C. Most of the following have a flash point much lower than this:

petrol	spirit varnish
cellulose thinner	polyurethane thinner
cellulose paint	methylated spirit
certain adhesives	chlorinated rubber thinner
white spirit	

The storage and use of these materials is controlled by the Highly Flammable Liquids and Liquefied Petroleum Gases Regulations (see Table 6.2).

Safety precautions
 (i) Do not handle near a naked flame, unprotected electrical equipment or source of spark.
 (ii) Take from store only the amount required for immediate use.
 (iii) Do not store in open or plastic containers.
 (iv) Replace screw tops as quickly as possible.
 (v) Return empty containers to store.

Other flammable liquids
Although not as volatile as petroleum spirits, these materials may produce an explosive vapour:

paraffin (kerosene)	creosote
oil paint	bitumen paint

Safety precautions The use and storage of these materials is not governed by such strict statutory regulations as apply to highly flammable liquids, but if the same rules are observed there will be less fire risk.

Gas stored in cylinders
(a) Butane, propane, acetylene
(b) Oxygen

The storage and use of these gases is controlled by the Highly Flammable Liquids and Liquefied Petroleum Gases Regulations and the Explosives (Acetylene) Order.

Safety precautions
(a) (i) Avoid sparks or naked lights, which are dangerous where 'free' gases are present.
 (ii) Do not use a hose which leaks or which can be easily creased.
 (iii) Ensure that connections to hoses, cylinder and burners are correct, tight-fitting and not cross-threaded.
 (iv) Keep flame away from cylinder and hoses.
 (v) Keep cylinder upright and do not knock violently.
 (vi) If acetylene cylinder becomes heated accidentally, shut valve, detach fittings and take into open air. Immerse in or apply water to cool; open valve till empty.
 (vii) Always ensure that acetylene regulator controls are turned off. Never open valves with the regulator on.
(b) The above rules also apply to oxygen. In addition, special precautions are necessary because, while oxygen does not burn, it supports and accelerates combustion. Oil and oil-impregnated clothing can be ignited by a spark and will burn fiercely in the presence of a concentrated supply of oxygen.
 (i) Keep fittings free from grease and oil.
 (ii) Do not use with greasy hands or clothing.
 (iii) Keep oxygen cylinders away from other cylinders.

Safe storage
 (i) When not in use, store in fire-resisting building or in a compound away from fire risk.

(ii) Store empty and full cylinders separately.

(iii) Mark empty cylinders clearly.

(iv) Close all valves tightly.

Flame-producing apparatus

Blow-lamps, gas torches and flame-cleaning torches may be dangerous unless used with great care in the vicinity of combustible materials.

Safety precautions

(i) Remove all litter and rubbish from area.

(ii) Protect combustible materials (e.g. timber floors) with metal or fire-proof sheeting.

(iii) Never leave lamps or torches burning when not in use.

(iv) Inspect area carefully after use to make sure that no smouldering material remains.

(v) Remove dark goggles regularly to inspect area to make sure that no surface is smouldering. (See also 1.1.4.)

Spark-producing apparatus

When working in an atmosphere which may be laden with flammable vapour or gas (e.g. when spraying oil or cellulose paint, in spray booths, in or near empty or full spirit/gas storage tanks) great care must be taken to avoid causing a spark which could ignite it.

Safety precautions

(i) Use spark-proof electric motors.

(ii) Use phosphor bronze or berylium copper tools, as steel may cause a spark when scraped on metal or stone.

(iii) Avoid clothing of man-made fibre, which can cause a spark due to static electricity.

(iv) Hoses for airless spray and abrasive cleaning must be earthed to avoid static electricity.

Spontaneous combustion

Some materials used by the painter do not require a spark or flame for ignition. Oil-soaked rag, if screwed up and left in a heap or a bin, and paint-soaked foam rubber, if placed in an enclosed bin, may smoulder and burn.

TABLE 6.1　Fire extinguishers

Type	Average range	Usual colour	Application	Prohibited use
Water or soda acid	4 m	Red	Wood Paper Rags Blow-lamps	Electrical fires Oil Solvents
Foam	4 m	Cream	Liquids Oil Paint	Electrical fires
Vapour-forming or vaporising liquid (BCF)	4 m	Green	Electrical fires Petrol or diesel engines	Fires in confined places
Vapour-forming gas (CO_2)	1 m	Black	Electrical fires Highly flammable liquids	Material containing own oxygen supply, e.g. cellulose
Dry powder	2 m	Blue	Highly flammable liquids Electrical fires Petrol or diesel engines	
Sand			Small isolated fires	
Glass-fibre blanket			Burning clothes Small isolated fires	
Aerosol (many types available)	Short		Small fires	

6.2.2 FIRE-FIGHTING (See Table 6.1)

A fire requires:

(a) *fuel* – combustible material which may be solid (e.g. wood), liquid (e.g. oil) or gas (e.g. petroleum vapour);

(b) *oxygen* – the flame requires oxygen to maintain it;

(c) *heat* – most solids and liquids require to be heated before they emit flammable vapours.

If any *one* of these factors is removed, burning will cease. The three types of fire most commonly met by the painter are fought in different ways:

(a) Solid fuel fires (wood, paper, cloth or rubbish): controlled by cooling with water.

(b) Liquid or gas fires (paint, oil or solvents): controlled by blanketing, using foam, powder or gas, to cut off the oxygen supply.

(c) Fires in electrical equipment (motors, wiring or switches): controlled by blanketing using non-conducting materials

6.2.3 TREATMENT FOR BURNING CLOTHING

(i) Lay the person down to keep flames away from face.

(ii) Quench the flames with water or other non-flammable liquid.

(iii) Wrap the person tightly in a rug, blanket or coat until the flames are smothered (nylon or other synthetic material must not be used).

If alone, roll on the floor and smother flames with the nearest available wrap; do not run about. (See 6.8.2.)

6.3 STORAGE OF MATERIALS

Any materials and tools kept in stock represent a large amount of money. Careless storage can prove to be expensive and, in some instances, dangerous.

Temperature Wherever practical the temperature of a storeroom should be kept constant. The ideal general temperature is 15°C, but because of the risk of evaporation and the resultant dangerous vapour, solvents should be stored at a lower temperature.

Dampness Many materials and brushes deteriorate when subjected to damp. A heated storeroom as described above rarely offers a degree of dampness that would affect materials.

Lighting Adequate natural and artificial lighting is essential for easy identification of stock and to avoid hazards. Care must be taken to ensure that all electrical wiring and fittings used for lighting are in good condition, otherwise they may become fire risks.

Fire precautions Very few painters' materials do not present a fire risk, therefore every precaution must be taken to keep this to a minimum by:

 (i) avoiding naked lights;
 (ii) banning smoking;
(iii) ensuring that correct types of fire extinguisher are available and in good condition.

Materials that come within the scope of the Highly Flammable Liquids and Liquefied Petroleum Gases Regulations must be stored under strict conditions as specified in Table 6.2.

TABLE 6.2 **Storage of materials**

Material	Type of storage	Special considerations
Oil paint Alkyd paint Polyurethane paint Oil varnish Polyurethane varnish Scumble Oil filler Spirit varnish Putty Bitumen	On shelves Clearly marked Arranged with new stock at back to avoid deterioration of old stock	Keep lids tightly closed to avoid evaporation and skinning Even temperature will ensure a reliable consistency Keep heavy containers at low levels to avoid difficult lifting Tins of heavily pigmented materials should be inverted at regular intervals to prevent settlement
Emulsion paint Emulsion varnish Acrylic paint Prepared paste Multi-colour paint	On shelves Clearly marked New stock placed behind existing stock	Protect from frost Many water-carried paints have limited shelf-life – it is essential that all stock is used within the time limit
Whiting Dry pigment Plaster Size Paste powder Powder filler	Small items on shelves Larger items on platforms at ground level Loose materials in closed bins	Protect from moisture Plasters have a limited shelf-life and may 'air set' if humidity or condensation is high

TABLE 6.2 **Storage of materials** – *continued*

Material	Type of storage	Special considerations
Spirit paint remover	On shelves	Temperatures in excess of 15°C may cause spirits to expand and blow out containers Avoid naked flames
Wallpaper Lining paper	In racks with ends protected Stacked by pattern and clearly marked	Keep wrapped to protect from dust and direct sunlight which may cause fading
Abrasive paper	In packets or sleeves for easy identification and to keep sheets flat	Avoid excessive heat which causes brittleness Avoid dampness which weakens glasspapers and garnet papers
Glass	In vertical racks	Requires dry storage to prevent sheets sticking together Dirty storage conditions can discolour glass (see 9.2.2)
Dust sheets	Folded On shelves	Keep clean and dry to avoid mildew
Brushes	Suspended or laid flat in cupboards New brushes should be kept wrapped	Protect against moth attack by using insecticide Keep dry to avoid mildew
Rollers	Suspend in cupboards	Lambswool and mohair should be treated as brushes
Steel tools Sprayguns	Suspended or laid flat in cupboards	Protect from corrosion by oiling or using anti-rust paper wrapping
Paraffin Creosote	(a) In large drums on trestles, tapped and locked (b) In 5 or 20 litre screw-topped cans at low level	Keep tightly stoppered Store outside in enclosed area, or in a building separate from main building (see also 6.1)
Liquefied gas Compressed gas Petrol Cellulose paint Cellulose thinner Chlorinated rubber paint and thinner Methylated spirit Polyurethane thinner Adhesives containing petroleum spirit White spirit	(a) In the open, protected from ice, snow and direct sunlight (b) In a special store, constructed as follows: With sloping floor, so that spilt solvent does not remain under its container Of brick, blocks, concrete or other fireproof material With roof of easily shattered material (asbestos) to minimise effect of explosion With doors 50 mm thick, opening outwards	Storage governed by the Highly Flammable Liquids and Liquefied Petroleum Gases Regulations *Note* These regulations apply only when more than 50 litres of material are stored Permission for storage must be obtained from the District Inspector of Factories

TABLE 6.2 **Storage of materials** – *continued*

Material	Type of storage	Special considerations
	With all glass wired, and not less than 6 mm thick	
	With concrete floors	
	With storage area surrounded by cill deep enough to contain contents of largest container	
	Lit with flameproof lights switched from outside	
	Unheated	
	Naturally ventilated at both high and low levels	
	Prohibited to naked lights or spark-production objects	
	Equipped with two or more exits	
	Prominently marked with red and white squares and a sign stating 'Highly flammable'	

6.4 LEAD PAINT

6.4.1 CONTROL OF LEAD AT WORK REGULATIONS 1980

These regulations came into force in 1981 and replaced the Lead Paint Regulations, 1927. They were prepared to protect any workers who may breathe in or swallow any material which may contain lead. The risk to the painter when applying coatings is very small because few paints today contain lead. The biggest danger to the painter is when old surface coatings are being prepared for redecoration. Many paints used in the first sixty years of this century contained lead pigments. When these are scraped or dry-abraded a dust will be released which could cause lead poisoning if inhaled or swallowed.

The regulations place equal responsibilities on the employer and the employee so that they both work and behave on sites in such a manner that risks of lead poisoning are extremely small.

Briefly the regulations cover the following points and apply to any site on which workers may be exposed to lead:

(i) The employer must provide similar protection to both his own employees and other people who may be working on the same site. This means that in occupied buildings like offices, schools, factories, every precaution must be taken to prevent lead dust being allowed to travel freely through the site.

(ii) Employers must inform employees of the risks to which they may be exposed. They must also tell them how to protect themselves against the possible risks.

(iii) Employers must ensure that no methods of preparation are carried out which will endanger the health of the employees. Specifying wet abrading reduces the risk of breathing or swallowing lead dust.

(iv) If the making of lead dust cannot be avoided employees must be provided with adequate respiratory protective equipment.

(v) Employers must provide protective clothing to all employees.

(vi) Washing facilities, and places to store protective clothing and personal clothing not worn during working hours, must be provided.

(vii) Smoking, eating and drinking are not permitted except in a place provided which is free from any danger of lead dust.

(viii) Every care must be taken to prevent spreading the risk of lead to other places. This requires proper cleaning of the site, and of protective clothing and respiratory equipment. Also ensuring that clothes worn on the site are not used elsewhere.

(ix) If the exposure to lead is particularly high employees must be examined by a doctor and strict records kept of the results.

Note Paints containing lead shall not be applied by spray in the interior of buildings.

6.4.2 REMOVAL OF PAINT SYSTEMS CONTAINING LEAD

All work involving the disturbance of coatings containing lead must be treated as potentially dangerous.

Where old coatings have to be removed:

(i) Determine whether the paint contains lead by using a simple test kit available from most trade merchants.

(ii) If lead is present use a suitable liquid – see 3.3.1.

Only in exceptional circumstance should lead paint be burnt off because if the nozzle temperature of the heat gun exceeds 450°C dangerous fumes and dust can be released. *Note* Most naked flame torches exceed 450°C. All softened paint must be scraped directly into a suitable container and disposed of in accordance with Waste Management Licensing Regulations 1994.

6.5 HYGIENE

6.5.1 GENERAL PRINCIPLES OF HYGIENE

It is an essential feature of good painting practice to work only in clean conditions. Such principles must also be applied personally. Most of the following points refer to site conditions required by the Construction (Health, Safety and Welfare) Regulations 1996.

Hand-washing must be carried out regularly, especially before meals. The regulations state that washing facilities must be provided on any site where people are employed for more than four hours.

The swallowing of old paint-film dust, paint particles, fungicide washes and pastes and many industrial dusts can have serious effects on the body, particularly the bowels.

Overalls or protective clothing must be worn for the entire period of work. The regulations require that accommodation for personal clothing must be provided on every site and this should be used to store outdoor clothing while overalls are worn on site.

Overalls should be washed regularly, at least once a week.

Wearing of overalls off the site is anti-social: it spreads the dirt and debris of the site and subjects other people to discomfort, irritation and the risk of disease.

Site facilities to comply with the regulations must include:
 (i) a place that offers protection during bad weather, in which warming and drying facilities are provided;
 (ii) a messroom in which there are arrangements for boiling water and heating food;
 (iii) sufficient drinking-water points that are easily accessible.

Sanitary conveniences must be provided which are reasonably accessible to all work places. They must be covered, partitioned to give privacy, have a door with a fastener, be lit and well-ventilated.

6.5.2 INDUSTRIAL DERMATITIS (OCCUPATIONAL SKIN DISEASE)

Dermatitis affects people in different ways: in a mild form it may result in slight irritation or redness of the skin; in more severe dermatitis the skin may harden, blister or crack, becoming exposed to a variety of infections. Warts and ulcers may result from severe skin damage. The disease does not necessarily show itself immediately: lack of care in handling harmful materials when young may result in skin disease later in life.

Ways in which the skin may be affected are shown in Table 6.3.

Prevention
The hands are the principal part of the body to be affected and there are various ways by which they can be protected.

Industrial gloves   Impervious gloves made from rubber, PVC or an impregnated material. Available in a variety of thicknesses and sizes. May cause severe perspiration which can have a detrimental effect on some skins, but this may be avoided by wearing thin cotton inner gloves. Gloves can be worn when working with any of the chemically active materials listed in Table 6.3.

Barrier cream   Only effective for short periods. Rubbed into every part of the hands *before* commencing work. There are two principal types:
(a) *Water-soluble*, which protects hands against attack by solvents and solvent-carried paints and resins. After work the hands are washed in water to remove all traces of cream and paint.
(b) *Water-repellent*, which is less common but essential when using water-based materials such as detergent or cement. Removed with skin cleanser, *not* solvent.

Skin cleansers are special blends of solvent oils, antiseptics and emulsifiers which remove paint, resins and barrier cream from hands without dissolving the natural oils in the skin. Available in liquid, jelly or paste form.

Note   The use of any solvent or detergent to clean hands is a dangerous practice which may cause dermatitis.

TABLE 6.3   **Causes of industrial dermatitis**

Cause	*Agent*
Chemical action (a) Irritation	Glass fibre Cement Epoxy, polyurethane, polyester, formaldehyde resins and hardeners Tar, pitch, bitumen Fungicide, insecticide Acid Alkali Certain dyes Certain glues
(b) De-fatting: dissolving of the protective fat or oil in the skin, leaving it open to infection	White spirit Paraffin Cellulose thinner Polyurethane, epoxy and chlorinated rubber paint thinners Methylated spirit Detergent Petrol Liquid paint remover
Obstruction of skin pores	Dirt Oil Grease (*Note* Also cause pimples and boils)
Mechanical damage: piercing of outer skin	Grit and dirt
Softening of the skin, with reduced protection against infection	Immersion in oils or watery solutions
Bacterial infection	Air-carried organisms trapped in infected skin

6.6 FUMES, DUST, NOISE, VIBRATION

6.6.1 FUMES

Asphyxia is one of the most common causes of unconsciousness. It occurs when a person is unable to breathe sufficient oxygen.

Asphyxia presents a treble danger:

(a) Becoming unconscious while on a scaffold can prove fatal.

(b) If the brain is deprived of oxygen for over four minutes it can be permanently damaged.

(c) Prolonged oxygen starvation causes death.

Solvent fumes

The space occupied by the vapour of a solvent is many thousand times that of the liquid. As spirit-carried paint dries, its solvent evaporates and, if unable to disperse, gradually replaces the air content of the space in which the work is being carried out. In a short time the painter is breathing spirit fumes which do not contain the oxygen that the body needs.

Materials that produce this kind of vapour are:

    oil paint

    alkyd paint

    polyurethane, epoxy and chlorinated
      rubber paint

    cellulose paint

    bitumen paint

    spirit paint remover

    oil and spirit varnish.

When spraying solvent-thinned materials in enclosed spaces the fumes can build up into an explosive mixture. Special extraction fans and trunking may be necessary if open doors and windows may release the fumes into the building and cause further hazards.

Gas

Unbreathable atmospheres can be produced by other forms of gas.

Carbon monoxide Exhaust gas from petrol or diesel engines progressively replaces oxygen in the blood, with lethal results.

Town gas Has similar effect to carbon monoxide.

Natural gas
Propane
Butane
    Non-poisonous, but if breathed in for sufficient length of time asphyxia will occur.

Smoke from any burning material gradually replaces oxygen (smoke is the cause of death more often than the actual fire).

Isocyanate A chemical used in polyurethane coatings and foams whose vapour can cause irreversible damage to lungs.

Chlorinated hydrocarbon solvents

As well as having the properties described for solvents above, these materials present another danger: in vapour form they decompose if exposed to a naked light or flame and form an irritant gas.

    Chlorinated hydrocarbon solvents are used in:

    spirit paint remover

    some cellulose paints

    some polyurethane and epoxy paints.

Paint-burning

Any form of combustion requires oxygen, therefore prolonged burning-off in a confined space will quickly use up the atmospheric oxygen. Some paints, e.g. polyurethane, also produce poisonous gases when burning (see 6.4.2).

Safety precautions

(i) The Construction (Health, Safety and Welfare) Regulations 1996 require that when operatives are working in the presence of injurious fumes either:

    (a) adequate ventilation must be provided, or

    (b) suitable respirators must be supplied.

(ii) When working inside a building with any of the solvents or paints listed above, ensure ample ventilation by opening windows and doors.

(iii) If ventilation is not possible, work in short spells only, going into the open air regularly to ensure adequate oxygen intake.

(iv) In very confined spaces such as storage tanks, pipelines, tunnels, breathing apparatus may be necessary: the use of this equipment requires special training. (See Table 6.4.)

(v) Do not smoke, use burning equipment, or allow any naked flame in the area, particularly when using chlorinated hydrocarbon solvents.

(vi) Ensure that petrol and diesel engines do not emit their exhaust into the interior of a building (this is also a requirement of the regulations quoted above).

(vii) Check that no hose connections, gas bottle or gas tap is leaking.

Note

(i) Face masks and respirators are of no protection when working in low oxygen atmospheres. They only clean the air being breathed, they do not replace the oxygen. Only breathing apparatus is effective under these conditions.

(ii) Many of the gases and vapours mentioned are heavier than air, therefore working at a low level in a room will prove more dangerous than in a standing or elevated position.

6.6.2 DUST

Unrestricted inhalation of dust can cause considerable damage to the respiratory organs. The damage is apparent as:

running nose and eyes
sore throat
headache and dizziness
inflammation of the lungs, which can lead to bronchitis.

Dust and grit can also severely damage the eyes unless protective equipment is worn.

Asbestos

The Control of Asbestos at Work Regulations 1985 control the use of asbestos products because serious lung disease (asbestosis) can be caused by inhaling the dust. The painter is generally concerned with asbestos cement whose effect is less severe than that of asbestos lagging, asbestos insulation boards or sprayed asbestos. Precautions should, however, be taken when producing any asbestos dust.

Stone, brick and concrete

Cleaning or abrading by mechanical tools, or grit-blasting, may release siliceous dust which if inhaled can cause severe damage to the lungs (silicosis). All these dusts can severely damage the eyes.

Timber

Excessive inhalation of any timber dust will cause respiratory damage but certain hardwoods, such as teak, obeche or iroko, and impregnated timbers, contain elements which can cause greater damage. Protection is essential when mechanically abrading timber in confined spaces or indoors.

Plastics

Plastic dust can cause lung damage. Glass fibre dusts also severely irritate the skin and damage internal organs if inhaled.

Safety precautions

(i) The Construction (Health, Safety and Welfare) Regulations 1996 require suitable respirators to be provided where operatives are required to work in dust-laden atmospheres.

(ii) The Protection of Eyes Regulations 1974 and/or the Personal Protective Equipment at Work Regulations 1992 require that:

(a) Suitable eye protectors or shields are provided where processes are carried out which could damage operatives' eyes.

(b) All persons provided with eye protectors MUST wear them while the conditions for which they are supplied exist.

(iii) Wherever possible use a wet-abrading process so that dust is not released.

(iv) Always wear a face mask or respirator when dust cannot be avoided. Breathing apparatus is necessary when asbestos dust is released.

(v) Remove excessive dust with an industrial vacuum cleaner, preferably one with High Efficiency Particle Air (HEPA) filter, rather than by brushing or sweeping.

6.6.3 NOISE

Prolonged or high level noise can cause tiredness, nausea or temporary or permanent deafness. Many mechanical tools used on construction sites can produce these symptoms.

Safety precautions The Noise at Work Regulations 1989 recommends:

TABLE 6.4 **Protective equipment**

	Equipment	Hazard	Notes
Respiration: prevention of inhalation of dust, fumes, spray, gas	*Mask* (Fig. 6.1) Disposable paper mask Disposable cotton filter pad	Large particles Non-toxic dust, e.g. paint spray, timber dust	Not safe for use in oxygen-deficient atmospheres
	Dust respirator (Fig. 6.2) Facepiece with replacement cotton/felt or impregnated wool filter	Fine or toxic dust, e.g. glass fibre, cement	Ditto
	Light fume respirator (Fig. 6.3) Facepiece with replaceable activated carbon filter	Fine dust and non-toxic fumes, e.g. spray	Ditto
	Breathing apparatus (Fig. 6.4) Orinasal facepiece or full-face mask or hood connected to a filtered air supply by a flexible air hose	Asphyxia in oxygen-deficient atmosphere, e.g. unventilated or very confined areas in which solvent-carried materials are used	Recommended for spraying two-pack epoxy and polyurethane coatings in enclosed spaces. Destroys sense of smell. Particularly dangerous when working with solvent-thinned materials as explosive atmospheres may build up
	Blast cleaning helmet (Fig. 6.5) Glass-fibre helmet with two-ply plastic window (outer can be replaced when badly abraded) and attached canvas cape covering shoulders, arms and chest. Can have air supply	Abrasive action of grit Breathing in highly concentrated dust atmosphere	
Eyes	*Impact and chemical goggles* (Fig. 6.6) (a) Clear or tinted acetate or polycarbonate lenses in tight-fitting PVC frames (b) Rubber or PVC frames with toughened glass lenses	Flying particles Liquid splashes Dust	Insufficient protection against glare or oxyacetylene flame
	Welders' goggles (Fig. 6.7) As above, with special blue or green filter glass or plastic lenses	Glare and flying particles from oxyacetylene torches	Allow only poor vision for normal use
Head	*Safety helmet or hard hat* (Fig. 6.8) Glass fibre or plastic, with soft padded headband and harness	Falling objects Collision with protrusions, scaffold poles or fittings	
Hands	*Industrial gloves* Leather, cotton twill Asbestos cloth Rubber, PVC, Neoprene or rubberised fabric	Abrasion, sharp edges Extreme heat Solvents, detergents	
	Industrial gauntlets Rubber, PVC or Neoprene Leather or rubber	Caustic soda Flying debris when grit-blasting	

TABLE 6.4 **Protective equipment** – *continued*

	Equipment	*Hazard*	*Notes*
Ears	*Earmuffs* (Fig. 6.9) Pair of padded plastic cups, either solid or liquid fillings, connected by simple head frame. Reduce high frequency noise but allow hearing of loud voices	Harmful noises at very high level, e.g. explosions, blasting, jet engines	Should be issued only after sound intensity has been measured and correct type of protection specified.
	Ear plugs or defenders Plastic or rubber plugs that fit into the ears and muffle noise. Various sizes available	As above	For use when ear muffs are not available
	Waxed cotton wool or glass fibre Moulded and fitted personally. They loosen with jaw movement, are less hygienic than plugs and more likely to cause infection	As above	For use when other forms of protection are not available
Feet	*Safety shoes or boots* With steel toe caps With steel sole lining (Many incorporate both features)	Dropping of heavy objects Stepping on sharp protrusions or objects	

Fig. 6.1 *Mask*

Fig. 6.4 *Breathing apparatus*

Fig. 6.7 *Welders' goggles*

Fig. 6.2 *Dust respirator*

Fig. 6.5 *Blast cleaning helmet*

Fig. 6.8 *Safety helmet*

Fig. 6.3 *Light fume respirator*

Fig. 6.6 *Goggles*

Fig. 6.9 *Earmuffs*

(a) If noise level is 90 dB(A) (e.g. within 20 m of a pneumatic drill) for more than 8 hours in a day, ear protectors should be worn.
(b) A noise level of 93 dB(A) should not be tolerated for more than 4 hours without protection.
(c) A noise level of 96 dB(A) should be tolerated for no more than 2 hours without protection.

Note dB(A) is a decibel scale which measures noise intensity. An increase of 3 dB(A) represents a doubling of sound intensity.

6.6.4 VIBRATION

Vibration produced by percussion hammers and needle guns used for long periods can cause operatives to lose all sensation in their hands: the condition is called 'white finger' or 'dead hand' (medical name 'Raynaud's phenomenon'). There is no medical remedy for this disease; once established, attacks may occur at any time, most often when not using vibrating tools and when the hands are cold.

Apart from the danger of dropped tools, it may make it possible for a man to injure his hand without feeling it.

Safety precautions
  (i) Restrict work to short spells.
 (ii) Wear gloves: keep hands warm.
(iii) Grip tool as lightly as possible.
(iv) Once symptoms are suspected, stop work immediately and seek medical advice.

6.6.5 RADIATION

Painters working in nuclear power stations can be exposed to ionising radiation.

Safety precautions A health register is required to be kept, in which exposure is recorded until the permissible maximum is reached or until a period of 30 years has elapsed. Records must be passed to any new employer.

6.7 LIFTING

During the normal course of his work the painter is required to lift heavy objects. If this lifting is carried out incorrectly it can cause injury. The Construction (Health, Safety and Welfare) Regulations 1996 and/or the Manual Handling Operations Regulations 1992 state that a person must not lift or carry any load which may cause an injury. Bad handling does not necessarily cause immediately noticeable injury, but may be responsible for the later development of rheumatism, fibrositis or slipped disc. Rupture or hernia is another serious injury which can be caused by incorrect lifting.

Correct handling and lifting also reduce effort and fatigue.

The *basic principles of handling and lifting* are as follows (see Fig. 6.10):

Keep back straight If the back is bent while lifting, the spinal column may be damaged. The safe method is to keep the back straight throughout the lift.

Bend the knees Crouch to the object with knees bent and back straight.

Tuck chin in With the head raised and the chin tucked in, a straight back is maintained.

Position of feet One foot should be placed behind and the other alongside the object, pointing in the direction of movement after the lift.

Firm grip Finger-tip grip is unsafe. The roots of the fingers and palm of the hand must be fully used to ensure a safe grip.

Arms close to the body If arms are spread out, the weight is transferred to fingers, arms and back, which could result in injury. The body takes the weight if the arms are kept close while lifting and carrying.

Smooth action Push off with rear foot, straighten the legs and move off in the required direction in one smooth movement.

Note Awkward-shaped objects require special care. When more than one person is lifting, appoint a leader to ensure that the lifting action is carried out together. If lifting gear is available, always use it.

Fig. 6.10 *Lifting*

6.8 FIRST AID

Safety rules and regulations are made to prevent accidents but because the nature of the work and the materials used by the painter are potentially hazardous, precautions must be taken to deal effectively and efficiently with an accident should it occur.

6.8.1 FIRST-AID REQUIREMENTS

The Health and Safety (First Aid) Regulations 1981 lay down the following for the provision of first-aid boxes and suitable first-aider support.

The regulations require that employers and self-employed persons provide adequate and appropriate first-aid facilities which should be based on an assessment of the need for first-aid depending on the nature of the work and *not just on number of people on site.*

First-aid boxes and kits
On small sites, where no special or unusual hazards exist and normal emergency services are readily available, the appointment of sufficient suitable persons trained in first-aid and the provision of first-aid boxes, with guidance and the presence of first-aid notices, may be adequate. Sites which present a high risk from hazards should be provided with a first-aid room under the responsibility of a suitable person with a suitable person available at all times. A suitable person is a first-aider who holds a current first-aid certificate issued by an organisation whose training and qualifications are approved by the Health and Safety Executive. Practising registered medical practitioners and nurses are considered as suitable persons.

First-aid boxes and travelling first-aid kits should contain a sufficient quantity of suitable first-aid materials *and nothing else.* In most cases these will be:
- 1 guidance card
- 20 individually wrapped sterile adhesive dressings
- 2 sterile eye pads

- 6 individually wrapped triangular bandages
- 6 safety pins
- 6 medium-sized individually wrapped sterile unmedicated wound dressings (approx. 100 mm × 80 mm)
- 2 large individually wrapped sterile unmedicated wound dressings (approx. 130 mm × 90 mm)
- 3 extra large individually wrapped sterile unmedicated wound dressings (approx. 280 mm × 180 mm)

Where mains tap water is not available for eye irrigation, sterile water or sterile normal saline (0.9%) in sealed disposable containers should be provided. *Eye baths/cups/refillable containers should not be used for eye irrigation.*

6.8.2 FIRST-AID TREATMENT

No matter how good the first-aid provisions on a site, there may always occur the situation when immediate positive action has to be taken, before trained help can arrive, to save the life or reduce unnecessary suffering of a workmate after an accident.

The following notes indicate first-aid treatment that an *untrained* person may be able to give in an *emergency* while awaiting the qualified aid or determining the extent of the injury. The treatment is not meant to be applied as an alternative to obtaining medical attention.

Burns and scalds
Caused by:
Dry heat Fire, flame, hot objects, sun
Electricity Electric current
Friction Revolving wheel, rope or wire
Corrosive chemicals Acids: sulphuric, hydrochloric, hydrofluoric, carbolic. Alkali: caustic, ammonia, lime.
Scalds Moist heat: boiling water, steam, hot oil or tar
Treatment
  (i) Reduce local effect of heat under slow

running water, or by immersing in cool water until pain has ceased (may be ten minutes).

(ii) Remove constricting objects, e.g. rings, belts, boots.

(iii) Remove soaked clothing but not burnt clothing, which is sterilised.

If the effect is severe:

(iv) Lay patient down.

(v) Cover burn with dressing or clean cloth.

(vi) Give small drinks at frequent intervals.

(vii) Arrange immediate removal to hospital.

Eye burn

Caused by corrosive chemicals.

Treatment

(i) Hold head under gently running tap; or plunge head into bucket or basin of water, instructing patient to blink continuously; or sit or lay patient with head tilted and flush copiously with cold or tepid water, milk, or sterile water from an approved EYE STATION if available.

(ii) Apply dressing or clean cloth.

(iii) Arrange immediate removal to hospital.

Minor wounds

Slight bleeding Wash, if possible with running water; apply dressing with pad and bandage.

Severe bleeding Stop bleeding by applying finger pressure to wound, either direct or over a dressing if immediately available; lay patient down and make comfortable; arrange for hospital treatment.

Bleeding from head Apply large dressing and fix with bandage; *do not press wound* in case skull is fractured.

Nose bleeding Support patient in sitting position with head slightly forward; ensure that the patient is breathing through mouth; instruct patient to pinch the soft part of the nose firmly for about ten minutes.

Discharge from ear Place and secure pad over ear; lay patient down with head slightly raised and inclined to side of injury; obtain medical advice.

Bruising Apply ice bag or damp material to reduce swelling and relieve pain.

Fractures

Limb fracture Steady and support injured part

to prevent further damage; immobilise by bandaging arm to body or leg to uninjured one; arrange immediate removal to hospital.

Skull fracture Symptoms: blood or fluid discharge from ear, vomiting, bloodshot eyes.

Treatment Place patient in recovery position (Fig. 6.11); apply sterile dressing to an ear discharge; keep watch on breathing; start artificial respiration if breathing stops; obtain urgent medical help.

Fig. 6.11 *Recovery position*

Jaw fracture Ensure airways not obstructed by tongue or blood; support jaw by hand or bandage; if severely injured place in recovery position; obtain urgent removal to hospital.

Severe back or spinal injury Do not move patient – warn him to lie still; cover with blanket or coat; arrange urgent removal to hospital.

Eye damage

All eye injuries are potentially serious and should receive immediate medical aid.

Injury to eyeball Foreign matter on eyeball or pupil, cause of damage unseen: close the eyelid; cover with large soft pad; arrange removal to hospital.

Foreign matter which is visible and *not* on pupil or adherent to eyeball: pull down lower lid or raise upper lid; remove matter with corner of clean handkerchief or small piece of damp cotton wool.

Unconsciousness due to asphyxia (See 6.6)

Caused by:

swelling of throat due to scalding or swallowing corrosives

drowning

poisoning by industrial gases (phosgene, solvent vapour, gas, carbon monoxide)

breathing smoke-laden or oxygen-deficient atmospheres.

Treatment
 (i) In the case of the last two causes, remove patient into open air or away from the affected area immediately.
 (ii) Remove any constricting garment, e.g. tie, buttoned collar, belt, waistband.
 (iii) Clear mouth of blood, vomit or swollen or swallowed tongue.
 (iv) If breathing is laboured or has stopped, apply respiratory resuscitation.

Severe electric shock
Caused by contact with low voltage (domestic supply) electrical current.
Treatment
 (i) Break the electrical contact by switching off current, removing the plug or wrenching the cable free. If these methods are not possible or cannot be carried out quickly, stand on dry insulating material (plastic, rubber) and, using dry wood, folded newspaper or rubber, push the patient's limbs away to break the contact.
Do not touch the patient with your hands.
 (ii) Treat burns as recommended earlier.
 (iii) If patient is unconscious, apply respiratory resuscitation immediately.

Methods of resuscitation
Mouth-to-mouth (Fig. 6.12)   Used if mouth or face is not damaged and there is no vomiting:
 (i) Pinch patient's nostrils together.
 (ii) Seal lips around mouth.
 (iii) Blow into lungs until chest rises.
 (iv) Remove mouth until chest falls.
 (v) Repeat at natural rate of breathing.
Silvester method (Fig. 6.13)   Used when face is damaged but chest and arms are not:
 (i) Lay patient on back.
 (ii) Raise the shoulders.
 (iii) Kneel astride the head.
 (iv) Grasp wrists and cross them over the lower part of the chest.
 (v) Rock body forward and press down on chest.
 (vi) Release pressure and with sweeping movement draw arms backwards and outwards as far as possible.
 (vii) Repeat rhythmically about 12 times a minute.
During recovery
After successful artificial resuscitation the injured person usually retches or vomits to clear the stomach of fluids. To prevent this causing choking or damaging the lungs the patient should be placed in the recovery position.

Fig. 6.12   *Mouth-to-mouth resuscitation*

Fig. 6.13   *Silvester method of
resuscitation*

PART SEVEN

Specification

7.1 Components of buildings – exterior 245

7.2 Components of buildings – interior 246

7.3 Components of buildings – staircases 247

7.4 Types of door and window 248

7.5 Job planning 250

7.5.1 Interior painting 250
7.5.2 Exterior painting 250
7.5.3 Wall and ceiling hanging 251
7.5.4 Construction documents 251

7.6 Choice of method of paint application 253

7.7 Costing 255

7.7.1 General procedure for large contracts 255
7.7.2 General procedure for small (jobbing) contracts 255
7.7.3 Information for costing small painting jobs 255
7.7.4 Glossary of terms 257

7.8 Calculations 258

7.8.1 International system of units (SI) 258
7.8.2 Calculating superficial area (area super) 258
7.8.3 Levers and beams 258

7.9 British Standards 261

7.9.1 British Standards 261
7.9.2 BS EN (European) Standards 262

7.1 COMPONENTS OF BUILDINGS – EXTERIOR (Fig. 7.1)

1 Cement flaunching	21 Hopper head
2 Chimney stack	22 Canopy
3 Hip tile	23 Bracket
4 Hip iron	24 Lintel
5 Soffit	25 Down pipe
6 Gutter	26 Fully-glazed door
7 Cladding	27 Threshold
8 Swan's neck	28 Gulley
9 Rainwater pipe (RWP)	29 Air brick
10 Bay window	30 Plinth
11 Shoe	31 Manhole cover
12 Ridge tile	32 Valley
13 Dormer window	33 Verge tile
14 Eaves tile	34 Barge board
15 Fascia board	35 Vent pipe
16 Vent light	36 Casement window
17 Dead light (picture window)	37 Sub light
18 Sill or cill	38 Gable
19 Waste pipe	39 Soil pipe
20 String course	40 Half-glazed door
	41 Step

Fig. 7.1   Components of buildings – exterior

7.2 COMPONENTS OF BUILDINGS – INTERIOR
(Figs. 7.2 and 7.3)

1 Ceiling rose
2 Cove cornice
3 Frieze
4 Picture rail
5 Wall filling
6 Niche
7 External mitre
8 External angle
9 Chimney breast
10 Mantle shelf
11 Return angle
12 Fire surround
13 Plinth block
14 Hearth
15 Internal mitre
16 Internal angle
17 Arch
18 Soffit
19 Architrave
20 Reveal
21 Sash window
22 Dado
23 Reeded column

Fig. 7.2   Components of buildings – interior

1 Ceiling bed
2 Pelmet
3 Bay window
4 Glazing bar
5 Apron panel
6 Skirting board
7 Cornice
8 Capital
9 Fanlight
10 Transom
11 Architrave
12 Pilaster (fluted or plain)
13 Wall panel
14 Panel moulding
15 Panelled door
16 Dado or chair rail
17 Dado panel
18 Plinth block
19 Margin

Fig. 7.3   Components of buildings – interior

7.3 COMPONENTS OF BUILDINGS – STAIRCASES (Fig. 7.4)

1 Skirting board
2 Hand rail
3 Landing
4 Nosing
5 Pendant end
6 Baluster
7 String capping
8 Wall string
9 Riser
10 Outer string
11 Spandrel panel
12 Newel post
13 Easing
14 Tread
15 Bull-nosed end

Fig. 7.4   Components of staircases

7.4 TYPES OF DOOR AND WINDOW
(Figs. 7.5 and 7.6)

flush   panelled   framed and braced   half glazed   fully glazed

1 Top rail
2 Panel moulding
3 Panel
4 Muntin
5 Stile
6 Middle or lock rail
7 Bottom rail
8 Finger plate
9 Door handle
10 Escutcheon
11 Kicking plate
12 Glazing bar
13 Solid panel
14 Weather board

Fig. 7.5   Components of doors

vertically sliding sash   vertically hung sash   horizontally hung sash   casement window

1 *Architrave*
2 *Soffit*
3 *Stop bead*
4 *Top (outer) sash*
5 *Meeting rail*
6 *Bottom (inner) sash*
7 *Pulley*
8 *Sash cord*
9 *Parting bead*
10 *Runner*
11 *External sill*

12 *Weathering*
13 *Ventlight*
14 *Transom*
15 *Hanging stile*
16 *Meeting stile*
17 *Rebates*
18 *Window frame*
19 *Mullion*
20 *Casement*
21 *Window board*

Fig. 7.6   *Components of windows*

7.5 JOB PLANNING

7.5.1 INTERIOR PAINTING

A typical method of organising the decoration of the room illustrated in Fig. 7.2 would be as follows:

Sequence of operations before painting

Remove carpets, pictures, mirrors and light fittings.
Remove small and light furniture.
Place heavy and bulky furniture in the centre of the room.
Cover floor and furniture with dustsheets.
Remove curtains and pelmets.
Remove all door and window furniture and fittings.

Sequence of operations for preparation

All processes should be carried out by starting at the highest point, and working downwards.
(i) Wash, abrade and make good surfaces in the following order:
Ceiling bed and cornice
Frieze, wall-filling and dado
Windows; replace defective glass and putties, and clean and straighten existing putties by scraping paint from glass
Doors and other woodwork.
(ii) Remove all rubbish from the site.
(iii) Spot-prime with suitable primer all bare patches exposed during the preparatory process.

Sequence of operations for painting

Working from top:
Ceiling bed and cornice
Frieze, wall-filling and dado
Windows
Doors
Other woodwork, completing with the skirting board.

Precautions when working

Always wear protective clothing.
Always use suitable plant and scaffolding.
Never stand on shelves, window boards or furniture.

Never rest stepladders on glass or glazing bars.

On completion

Clean all glass.
Clean and replace door and window furniture.
Replace carpets, curtains, pelmets and other furniture.

7.5.2 EXTERIOR PAINTING

To avoid working over clean finished work, all preparation and painting should be started at the highest level. A recommended method of working on the house illustrated in Fig. 7.1 is as follows.

Sequence of operations before painting

Remove or tuck away curtains, especially when burning off.
Lay protective sheets to protect paving and stonework.
Cover plants and shrubs near the building.

Sequence of operations for preparation

(i) Wash, abrade and make good surfaces in the following order:
Dormer window: also replace broken glass and defective putties
Barge boards
Gutters: also clean insides and repair defective joints
Fascia boards
Soffits
First floor windows: also replace broken glass and defective putties
Cladding
Ground floor window and cills: also replace broken glass and defective putties
Canopy
Brickwork or rendering
Rainwater, waste and soil pipes
Doors and other woodwork
(ii) Remove all rubbish from site.

(iii) Spot-prime all bare patches exposed during preparatory process.

Sequence of operations for painting

Working from the top, paint in the same order as for preparation.

Precautions when working

Always stand ladders on scaffold boards to prevent damage to lawns.

Always pad tops of ladders to prevent damage to freshly painted surfaces.

Always use a kettle-hook when working on ladders (see 1.1.2).

Never rest ladders on glass, glazing bars or gutters.

Never drag ladders down walls or across gutters.

On completion

Clean up paint spots.

Clean and replace door furniture and fittings.

Clean all glass.

7.5.3 WALL AND CEILING HANGING

Sequence of operations before hanging surface coverings

Prepare as for interior painting.

Undercoat as for interior painting, omitting surfaces to be papered.

Size and cross-line if necessary (see 4.1).

Complete painting as for interior painting, taking paint 6 mm on to lining paper at cornice, architraves and skirting.

Sequence of operations for papering

(i) *Walls*

Either (a) start at the window and work away from it in both directions or (b) work either side of centre line (see 4.10, Centring).

If chimney breast or other focal point exists, strike a centre line and work from either side.

If possible, remove or loosen switches and fittings as they are reached and fit paper behind before refixing.

Note Turn off power before loosening switches.

Do not fit foil wall vinyls behind switches.

Finish in a dark or unobtrusive corner where the loss of matching pattern will not be noticed.

(ii) *Ceilings*

Work from the light.

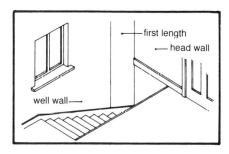

Fig. 7.7 *Papering a staircase well*

Hang paper across the room, working away from the window.

(iii) *Staircases* (Fig. 7.7)

Hang first length on well wall, at junction with head wall or apron lining.

Work away from the first length in both directions, finishing in a dark corner.

Precautions when working

Match and cut all lengths before pasting and hanging (with the exception of staircase work).

Fold up waste material and place under paste bench or in suitable container.

Remove paste from paintwork while still wet.

On completion

As for interior painting.

7.5.4 CONSTRUCTION DOCUMENTS

The planning and running of jobs of all sizes requires organisation and records. A few of the most important documents used for this purpose are as follows:

Architects' drawings

A range of scale working drawings which form the primary means of communication in the construction industry. Consist of six main types:

Block plan Shows the proposed building site in relation to the surrounding houses and roads, etc. Usually to a scale of 1:2500 and 1:1250.

Site plan Shows the position of the proposed building and the general layout of adjacent roads, services, trees and drainage, etc. Usually to a scale of 1:500 and 1:200.

Location drawings Show plans and elevations of the various areas within a building and the location of the principal elements of the building. Usually to a scale of 1:200, 1:100 and 1:50.

Range drawings Show the sizes of a standard range of building components such as kitchen units and windows, etc. Usually to a scale of 1:100, 1:50 and 1:20.

Detail drawings Show all the information required to manufacture particular components such as windows and doors, etc. Usually to a scale of 1:10, 1:5 and 1:1.

Assembly drawings Show details of the junctions between the various elements and components of a building such as between door frame and wall, etc. Usually to a scale of 1:20, 1:10 and 1:5.

Bill of quantities

A book of calculated quantities of materials, labour and other items for a proposed job. Compiled by the quantity surveyor and based on the architect's drawings, specifications and schedules.

Delivery notes

Paper record which accompanies all deliveries of materials and plant to the site. Someone is required to carefully check the items delivered compared to the details on the delivery note and sign that they have been delivered. Any items not delivered should be noted on the form. All notes should be sent to the employer to confirm that goods have been received before the employer pays invoices.

Progress charts

A visual chart on which a proposed job can be planned from start to finish ensuring that appropriate deliveries of materials and supply of labour can be organised at the right time in the job. The two main types used are the bar chart and the gant chart.

Requisition/order forms

Form on which to order plant and materials for a particular job. The order form number and delivery note for the goods delivered should match, as should the invoice sent to the employer for payment.

Schedules

These are used to record repetitive information, (size, colour, position, etc.) for a range of similar components such as paint finishes, doors, windows, etc. There can also be a schedule which lists all the drawings related to a job.

Schedule of rates

A list of prices per m^2 for specified tasks. They include all materials, labour, overheads and profit for the work described.

Specification

Contains precise descriptions of materials to be used and processes to be carried out which cannot be shown on the architect's drawings. Any variations to the specification can only be authorised by the agent. Also contains other information such as requirements for site clearance, making good on completion and who is responsible for approving finished work, etc.

Timesheets

An individual daily and weekly record of work done and time taken. Usually completed by the individual operative and verified by the supervisor.

Variation order

A form which authorises variations to the specification. They must be agreed and signed by the agent so that they can legitimately be used to change the final sum of the contract.

7.6 CHOICE OF METHOD OF PAINT APPLICATION (Table 7.1)

As paint technology has advanced, so alternative methods of application have been devised to cope with the new materials produced. The conventional brush has been superseded for some purposes by methods of application designed to overcome the problems involved in applying rapid-drying, high viscosity and textured coatings, while at the same time producing a high quality, even coating. In some cases an alternative method may simply be more productive than the brush.

TABLE 7.1 **Comparison of methods of paint application**

Method	Advantages	Disadvantages	Common use
Brush	Smooth finish Ability to cut-in Works paint into crevices to gain adhesion Essential for priming Ability to reach otherwise inaccessible places Very little splashing Economical with material	Slow and laborious on large or textured areas Unsuitable for rapid-drying paints Animal-hair fillings affected by caustic and liquid paint removers	All materials having a sufficiently long initial set Typical use – 25 mm: cutting-in glazing bars 50 mm: panelled door 75 mm: flush door 100 mm: ceilings and walls 150 mm: large areas
Paint roller	2–3 times faster than the brush on suitable surfaces Applies thicker coating than the brush No brush-marks Can be used fixed to extension pole	Splashes more than the brush Loose fibres may spoil the finish Inaccurate for cutting-in Cleaning can be time-consuming Paint containers cumbersome on scaffolding Textured finish not always acceptable	All materials having a sufficiently long initial set Large areas with the minimum of cutting-in Relief and textured areas Flat oil paints, avoiding the need to stipple Chain-link fencing and railings Acoustic tiles
Conventional spray	4–5 times faster than the brush on suitable surfaces No brush-marks Can reach almost inaccessible places High quality of finish Easy control of spray pattern	Produces overspray (paint in the air), adjacent work not required to be painted must be protected from overspray by masking Respirators and extractor fans may be required for inside use Uneconomical with paint and cleaning solvents Cleaning can be time-consuming Paints must be thinned to pass through the gun The preparation and maintenance of the equipment makes its use uneconomic for small jobs	Large unbroken areas Textured areas Rapid-drying paints Awkward, irregular shapes All paints that can be satisfactorily thinned and will flow Multi-colour paints

TABLE 7.1 **Comparison of methods of paint application** – *continued*

Method	Advantages	Disadvantages	Common use
Airless spray	Up to 30 times faster than the brush on suitable surfaces No brush-marks Very little or no overspray Large spray pattern Ability to cut-in within 15 mm Paint-thinning unnecessary Applies a thick coating with the minimum of strokes Paint flows out better than conventional spray No contamination from air line impurities	Equipment initially expensive Cleaning can be time-consuming In spite of precautions, guns tend to block-up, although reversible tips simplify the clearing of blocked fluid-tips High velocity of fluid jet presents a hazard at close quarters, especially when fluid-tip is removed Fluid-tip must be changed to alter the spray pattern	Very large areas: gas holders, ships, aircraft hangers Thick or thin paints Heavily textured surfaces Awkward and irregular shapes

7.7 COSTING

7.7.1 GENERAL PROCEDURE FOR LARGE CONTRACTS

(i) Architect or designer prepares a specification and drawings of the proposed job based on the client's requirements.
(ii) Quantity surveyor prepares a bill of quantities based on the specification and drawings.
(iii) Bill of quantities and drawings are sent out to decorating contractors with an invitation to submit a tender.
(iv) Decorating contractor costs the job and prepares the estimate from the bill of quantities and submits the tender.
(v) The accepted estimate becomes the price which the client will pay at the satisfactory completion of the job (plus or minus certain extras).

7.7.2 GENERAL PROCEDURE FOR SMALL (JOBBING) CONTRACTS

(i) Decorating contractor is approached by the prospective client.
(ii) Decorating contractor obtains the details of the job from the client or the client's agent.
(iii) Decorating contractor takes all necessary site measurements of the job.
(iv) Decorating contractor prepares a specification of the work to be done and an estimate of its cost which is submitted to the client for approval.

7.7.3 INFORMATION FOR COSTING SMALL PAINTING JOBS

A method of taking site measurements and preparing the specfication is as follows.

Job dimensions
Measuring-up on site should be carried out accurately with the use of 2 metre rod or stick, measuring tape or ordinary folding rule. Dimensions are then recorded in a Dim book, or on paper ruled in a similar way. Job details can be recorded at the same time as dimensions, e.g.

6/	0.750 m 2.000 m	9 m^2	
			2 oil paint on flush doors
2/	2.000 m 2.000 m	8 m^2	
			2 oil paint on casement windows plus 20% cutting-in

Column 1 Quantity of items measured (6 doors, 2 windows).
Column 2 Actual dimensions of the item, recorded from the top: *length* followed by *width* followed by *height*.
Column 3 Actual total area of the items measured.
Column 4 Description of the item and brief specification of the work to be done, often abbreviated, e.g.

Component	Existing condition	Work to be carried out
Ceiling and cornice	Papered and painted in sound condition	2 coats white emulsion
Walls	Papered in sound condition	Strip, cross-line, hang supplied paper

G & V   grain and varnish
K.P.S.   knot, prime and stop
E.G.   eaves, gutter
R.W.P.   rainwater pipe
2   two coats

Note the additional percentages to cover time-consuming work, e.g.

Cutting-in windows   10–50 per cent
Painting pipes   10–50 per cent
Painting corrugations   30–50 per cent
Painting gutters   10–50 per cent

Job details and specification

Notes should be made on the condition of the job, e.g. Is burning-off necessary? What is condition of wall beneath old wallpaper? Client's requirements of work should be recorded accurately room by room in the sequence of operations for internal painting (see 7.5).

Preparing the specification   As the specification is a binding contract, it must be written in a simple and clear manner so that there is no confusion as to what the client will be paying for, or how the work is to be done. The full preparatory process should be described, the material and colour to be used and the method of application.

Supplied specifications are usually laid out in the following order:
  (i) *Job headings:* brief description of the work, client's name and address, site address, architect.
 (ii) *Drawings:* copies of all necessary plans and elevations to assist in defining means of access.
(iii) *Preliminaries:* a list of general clauses which would otherwise appear many times in the specification, covering e.g. British Standards, quality of workmanship and materials, construction regulations.
(iv) *Clauses:* separate, explicit instructions and details of work to be carried out. Each page numbered and the clause lettered, e.g. Clause2/D (clause D on page 2).
 (v) *Appendices:* other relevant information, e.g. position of electrical power supply, water supply and rest rooms.

Costing the job

The price of a job consists of:

  (i) the total cost of labour for the job;
 (ii) the total cost of materials for the job;
(iii) an allowance for overheads for the period of the job;
(iv) an allowance for profit.

The following variables affect the costs involved, and they are used in different ways depending on the size and complexity of the job, and the administrative organisation of the contractor.

Labour hourly rate

The total cost of labour per employee hour. It is calculated not only on the employee's weekly wage, but also includes numerous other expenses such as travelling expenses, overall allowance, public and annual holiday payments, redundancy contributions, National Health and Insurance contributions.

The weekly total of all these expenses is divided by the standard number of hours worked per week to calculate the labour hourly rate.

Material cost

Material costs are supplied by the distributor and expressed as cost per litre. They may vary from month to month and current price-lists should always be consulted.

Spreading capacities of materials

These are supplied by the manufacturer and expressed as square metres per litre (m^2 per 1) and are calculated on normal brush application to surfaces of average absorption and texture. Spreading capacity can vary by up to 25 per cent either way, depending on method of application, texture, temperature and absorption. For spreading capacities of specific materials see Part Two: Surface Coatings.

Labour constant

The amount of work an employee can be expected to do in one hour, expressed as square metres per hour. It is dependent on the condition and type of surface, the material being used, the process involved and, most important, the method of application:

Process or material	m^2 per hour
Wash, abrade, make good walls	3–5
Gloss painting by brush	7

Cost of overheads

This will vary from one contractor to another depending on the size of the firm and the quality of the building and administration. Overheads cover such items as insurances, non-productive staff salaries, rent, rates, telephone, power, depreciation of tools, plant and equipment, stationery and advertising. Overheads, when coupled with profit, may be expressed as a percentage of the prime cost.

Profit

This will vary depending on the firm's overheads and estimated profit. It is expressed as a percentage ranging from 25 to 100 per cent which is added to the prime cost. Calculated to cover the cost of overheads and leave a net profit for the contractor.

7.7.4 GLOSSARY OF TERMS

Agent   Person acting on behalf of the client.

Architect   Designer of buildings. Specifies the materials to be used and the method of use. Often acts as agent on small sites.

Bill of quantities   A book of calculated quantities of materials for a proposed job, made up by a quantity surveyor, based on the architect's specification.

Client   Person or persons requiring and paying for work to be done. Usually the owner of the property.

Daywork rate   A unit rate per hour to cover labour and overhead costs. Often used as the basis for a provisional sum, or for work which is too complex to quote accurately.

Dim book   Small pocket book ruled in a special way in which to record dimensions when measuring work.

Estimate   An estimate of the cost of the job prepared by the contractor. It is not a binding contract, and allows for unforeseen additions which could not be seen under normal circumstances, such as the surface condition under existing wallpaper.

Estimated profit   The net profit that a contractor expects to make on an estimated job.

Estimated contract   A quotation which makes allowances in the final price for rising labour and material costs. The period of the contract is agreed between the client and the contractor.

Fixed price contract   An estimate which cannot under any circumstances be altered by subsequent rising costs of labour or materials.

Gross pay   Employee's pay before deductions such as income tax, insurance contributions and superannuation.

Gross profit   Profit made from a contract before the deductions to cover overheads.

Hourly rate   For costing purposes, the employee's hourly paid rate plus additions to cover the cost of overheads such as travelling, insurance and holidays.

Labour constant   The estimated and averaged amount of work one painter can do in one hour (m^2 per hour).

Net pay   Employee's pay after deductions of income tax, insurance contributions and superannuation. A painter's 'take home' pay.

Net profit   Profit made from a contract after the deductions to cover overheads. Employer pays income tax on this amount.

Overheads   The cost of running a business, such as non-productive staff salaries, rent and transport.

Prime cost   The actual cost of labour and materials before the addition of overheads and profit.

Provisional sum   A quotation for a separate quantity of work. Supplied for parts of a job which are too complex to accurately estimate time or cost. Often quoted on a daywork rate basis.

Quotation   A binding contract prepared by the contractor which shows the client the exact amount to be paid for a stipulated quantity, and quality of work. It can only be amended for additional work ordered by the client at an agreed rate.

Schedule of rates   A list of prices per m^2 for specified tasks. Includes all overheads and profit. Formulated from current material and labour costs.

Specification   A binding contract which defines in detail the work to be carried out, also colours and types of material to be used on a contract.

Tender   An estimate submitted to a client. In an *open tender* all contractors are invited, in a *closed tender*, only specified contractors.

7.8 CALCULATIONS

Fig. 7.8 *Calculation of surface area*

7.8.1 INTERNATIONAL SYSTEM OF UNITS (SI)

A metric system of measurement in which the principal units are usually divided or multiplied by one thousand to obtain the smaller or larger units. The prefix *milli* means a one thousandth part. The prefix *kilo* means one thousand times.

Correct writing of numbers

(i) Omit the comma and leave one space when writing numbers exceeding ten thousand, e.g. 32 000, 170 455, 11 060 (*but* 1000, 3240).

(ii) Arrange large numbers in groups of three, e.g. 302 270 355.

(iii) Express decimal fractions to three decimal places, e.g. 2.350 m, 0.750 m, 9.075 m.

(iv) Express fractions of metres either as decimals or as whole numbers of millimetres, e.g. 0.750 m = 750 mm, 0.065 m = 65 mm. (Also, £0.75 = 75p.)

(v) Give dimensions in order of length × width × height, e.g. a room measures 4.400 m × 3.050 m × 2.995 m.

Correct writing of symbols

(i) In giving dimensions, do not mix units, e.g. use 2.100 m × 0.750 m or 2100 mm × 750 mm (*not* 2.100 m × 750 mm).

(ii) Leave a space between numbers and symbol, e.g. 5.300 m.

(iii) Do not add a comma or a full stop after a symbol unless it is required for normal punctuation.

7.8.2 CALCULATING SUPERFICIAL AREA (AREA SUPER) (Table 7.3)

The surface area of a shape is calculated in squared units which represent the number of squares contained on the surface of the shape, e.g. a wall measuring 5.000 m × 3.000 m has a surface area of $5 \times 3 = 15 \text{ m}^2$ (Fig. 7.8).

7.8.3 LEVERS AND BEAMS

The main application of the principle of levers to painting and decorating is in the calculation of the counterweights required to secure safely a cradle or safety chair (see 5.4.1).

A further application is in the calculation of the loads carried by trestles (Fig. 7.9).

$$\text{Load } L = \frac{W \times A}{T} \text{ and load } R = \frac{W \times B}{T}$$

where T = length of board between supports
B = distance WL
A = distance WR
L = left load
R = right load
W = total imposed weight

Fig. 7.9 *Calculation of load*

In the example shown,

$$L = \frac{76 \times 1}{4} = \frac{76}{4} = 19 \text{ kg}$$

As the total load $W = 76$ kg and load $L = 19$ kg, load $R = 76 - 19 = 57$ kg. It will be seen that the support nearer the load bears the greater part of the weight.

TABLE 7.2 Principal units used in painting and decorating

QUANTITY MEASURED	UNIT	SYMBOL	SMALLER UNIT	SYMBOL	LARGER UNIT	SYMBOL
Length or linear measurement	metre	m	millimetre (m ÷ 1000)	mm	kilometre (m × 1000)	km
Area or superficial measurement	square metre	m^2	square millimetre (m^2 ÷ 1 000 000)	mm^2	hectare (m^2 × 10 000)	ha
Volume	cubic metre	m^3	cubic millimetre	mm^3		
Capacity	litre	l	millilitre (l ÷ 1000)	ml		
Weight	kilogram	kg	gram (kg ÷ 1000)	g	tonne (kg × 1000)	t
Pressure	bar	bar	millibar (bar ÷ 1000)	mbar		
Volume rate of flow	litres per second	l/s				
Temperature	Degree celsius	°C				
Precision measurement	Micrometre	μm			millimetre (μm × 1000)	mm

TABLE 7.3 **Calculating the surface areas of common shapes**

SHAPE	APPEARANCE	FORMULA FOR CALCULATING THE AREA
Square		Multiply length × height (l × h)
Rectangle		Multiply length × height (l × h)
Triangle		Multiply half length × perpendicular height ($\frac{1}{2}$ l × ph) or $\frac{l \times ph}{2}$
Circle		Multiply π (3.142) × radius × radius (πr^2) *Note* The circumference is calculated by multiplying π × diameter (π d)
Cylinder		Multiply π (3.142) × diameter × length (π dl) or circumference × length
Sphere		Multiply 4 × π × radius × radius ($4\pi r^2$) *Note* A hemisphere or dome is $2\pi r^2$
Walls of a room		Multiply the distance around the room (girth) × height of the room (girth × h)

7.9 BRITISH STANDARDS

The following British Standard specifications and British Standard EN (European) Standards may be referred to when writing or interpreting specifications. They also contain specifications for products to support safety legislation.

7.9.1 BRITISH STANDARDS

BS 144:1990 — *Wood preservation using coal tar creosotes*

BS 245:1992 — *Specification for mineral solvents for paints and other purposes*

BS 381c:1996 — *Specification for colours for identification, coding and special purposes*

BS 476 (Parts 1–33) — *Fire tests on building materials and structures*

BS 952 (Parts 1–2) — *Glass for glazing*

BS 1070:1993 — *Specification for black paint (tar-based)*

BS 1129:1990 — *Specification for portable timber ladders, steps, trestles and lightweight staging*

BS 1139 (Parts 1–5) — *Metal scaffolding*

BS 1191 (Parts 1–2) — *Specification for gypsum building plasters*

BS 1192 (Parts 1–5) — *Construction drawing practice*

BS 1248 (Parts 1–3) — *Wallcoverings in roll form*

BS 1336:1996 — *Specification for knotting*

BS 1710:1991 — *Specification for identication of pipelines and services*

BS 2015:1992 — *Glossary of paint and related terms*

BS 2037:1994 — *Specification for portable aluminium ladders, steps, trestles and lightweight staging*

BS 2482:1990 — *Specification for timber scaffold boards*

BS 2769:1984 — *Hand-held electric motor-operated tools*

BS 2830:1994 — *Specification for suspended access equipment for use in the building, engineering, construction, steeplejack and cleaning industries*

BS 2992:1987 — *Specification for painters' and decorators' brushes*

BS 3046:1991 — *Specification for adhesives for hanging flexible wallcoverings*

BS 3416:1991 — *Specification for bitumen-based coatings for cold application*

BS 3761:1995 — *Specification for solvent-based paint remover*

BS 3900 (Parts A–J) — *Methods of test for paints*

BS 3913:1989 — *Specification for industrial safety nets*

BS 4652:1995 — *Specification for zinc-rich priming paint (organic media)*

BS 4756:1991 — *Specification for ready-mixed aluminium priming paints for woodwork*

BS 4764:1991 — *Specification for powder cement paints*

BS 4800:1994 — *Schedule of paint colours for building purposes*

BS 5082:1993 — *Specification for water-borne priming paints for woodwork*

BS 5252 : 1996 — *Framework for colour co-ordination for building purposes*

BS 5358 : 1993 — *Specification for solvent-borne priming paints for woodwork*

BS 5378 (Parts 1–3) — *Safety signs and colours*

BS 5493 : 1977 — *Code of practice for protective coating of iron and steel structures against corrosion*

BS 5499 (Parts 1–3) — *Fire safety signs, notices and graphic symbols*

BS 5555 : 1993 — *Specification for SI units and recommendations for the use of their multiples and of certain other units*

BS 5589 : 1989 — *Code of practice for preservation of timber*

BS 5973 : 1993 — *Code of practice for access and working scaffolds and special scaffold structures in steel*

BS 5974 : 1990 — *Code of practice for temporarily installed suspended scaffolds and access equipment*

BS 6100 (Parts 1–6) — *Glossary of building and civil engineering terms*

BS 6150 : 1991 — *Code of practice for painting of buildings*

BS 6262 : 1982 — *Code of practice for glazing in buildings*

BS 7028 : 1988 — *Guide for selection, use and maintenance of eye-protection for industrial and other users*

BS 7079 (Parts A–F) — *Preparation of steel substrates before application of paints and related products*

BS 7664 : 1993 — *Specification for undercoats and finishing paints*

BS 7719 : 1994 — *Specification for water-borne emulsion paints for interior use*

BS 7779 : 1994 — *Specification for semi-transparent priming stains (stain basecoats) for joinery*

7.9.2 BS EN (EUROPEAN) STANDARDS

BS EN 131 : 1990 — *Ladders*

BS EN 266 : 1992 — *Specification for textile wallcoverings*

BS EN 354–365 : 1993 — *Personal protective equipment against falls from heights*

BS EN 971 : 1996 — *Paints and varnishes; terms and conditions for crating materials*

PART EIGHT
Colour

8.1 Colour and light   265

8.1.1 The spectrum   265
8.1.2 Effect of artificial light on colour   266
8.1.3 Types of artificial light   266
8.1.4 Metameric colour effect   266

8.2 Colour organisation   267

8.2.1 Pigment mixture colour circle   267
8.2.2 Munsell scale   267

8.3 Colour standards   270

8.3.1 BS 1710:1991 *Specification for identification of pipelines and services*   270
8.3.2 BS 4800:1994 *Schedule of paint colours for building purposes*   270
8.3.3 BS 5378:1995 (*Part 1*) *Safety signs and colours*   271

8.4 Terms related to colour schemes   274

8.1 COLOUR AND LIGHT

8.1.1 THE SPECTRUM

The source of all colour is light. Without light, colour does not exist. The purest form of light originates from the sun and is known as 'white light'. It is received by us in rays which are not straight but formed of waves of different lengths of frequencies. When light strikes a surface, waves of certain lengths are absorbed and others are reflected. It is the reflected waves which are picked up by our eyes and registered by the brain as colours.

The spectrum (Fig. 8.1)
When light is directed through a transparent material such as glass its speed is slowed down. If at the same time the light rays can be 'bent', e.g. through a glass prism, when the light emerges from the other side it will have been divided into individual colour waves. This can easily be seen if rays are projected on to a white screen. This coloured reflection will show a distinct range of colours always in the same order: red, orange, yellow, green, blue and violet, known as the spectrum. A similar effect is produced in nature when sunlight is broken up by raindrops and a rainbow is produced.

white light — red, orange, yellow, green, blue, violet

Fig. 8.1 *The spectrum*

Seeing colour (Fig. 8.2)
An object which is seen by a person with normal vision as red is absorbing all the light of other wavelengths and reflecting only the waves of 'red' length; the sensation caused by these waves meeting the eye is registered as red by the brain.

A white surface reflects all waves, whereas a black surface absorbs all, reflecting none. If the light source is not white the surface will appear a different colour: e.g. if red light is directed on to a green surface (green = blue + yellow), only a neutral or 'grey' colour will be seen because the light contains no waves of 'green' length to be reflected.

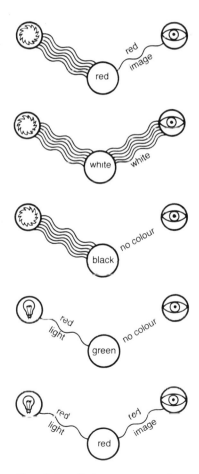

Fig. 8.2 *Colour perception*

8.1.2 EFFECT OF ARTIFICIAL LIGHT ON COLOUR

Colours seen in 'white' or natural light may appear different when seen in artificial light. This happens because the spectrum of artificial lights differ from that of white light, therefore the reflective rays picked up by our eyes will be different.

For this reason it is essential to work only in the main light source of a particular situation when selecting or mixing colours. Colours selected in natural light may look entirely different under artificial light.

8.1.3 TYPES OF ARTIFICIAL LIGHT

There are many forms of artificial light, but they can be grouped under three main sources which have common spectrums.

Tungsten Also known as filament lamps or normal domestic bulbs. The glass bulb contains a gas and the filament inside is made of thin tungsten wire which gives off a bright yellow/red glow when electricity passes through it. The light given off contains very few blue and violet rays.

Fluorescent Also called tube or strip lights. The tube contains low pressure mercury vapour which produces ultra-violet light when electricity passes through it. The inside of the tube is coated with fluorescent powders which glow when in contact with the ultra-violet light. The colour of the light depends on the type of powder in the tube. Most types of domestic tubes give out more blue/violet rays than red/yellow.

Discharge Similar to fluorescent tubes but without a coating of fluorescent powder. The vapour in the tube is at a much higher pressure and the colour of light emitted depends on the type of vapour. Sodium vapour gives off only yellow/amber light rays. Mercury vapour gives out violet, blue and green rays but very few red. Discharge lamps are used mainly for industrial and street lighting.

8.1.4 METAMERIC COLOUR EFFECT

Some dyes used in fabrics and paints may appear completely different when seen under various light sources. This is caused by the chemical properties of the dye which react to the various lights in different ways. This is another reason for only working in the relevant light source when selecting and mixing colours.

8.2 COLOUR ORGANISATION

Many methods of putting the known facts about a colour into a practical and usable order have been prepared. The two most useful to the painter and decorator are the pigment mixture colour circle (sometimes referred to as the Brewster theory) and the Munsell scale.

8.2.1 PIGMENT MIXTURE COLOUR CIRCLE (Fig. 8.3)

When the spectrum colours are arranged in a circle they are easier to use and to relate to each other. The pigment mixture circle consists of *three primary colours* and *three secondary colours*. The primary colours are yellow, red and blue; these are the basis of all other colours and cannot be produced by mixing other colours. The secondary colours are orange, purple and green and are produced by mixing together two primary colours. When the three primary colours are mixed together, neutral grey is the result.

Tertiary colours are mixtures of the secondary colours. Orange and purple; purple and green; and green and orange.

Note When colours are mixed together the resulting colour is not as bright or pure as the original colours. Therefore secondaries are less bright than primaries and tertiaries are less bright than secondaries.

8.2.2 MUNSELL SCALE (Fig. 8.4)

This is an international system which permits an accurate verbal description of colour. The method can be used:

(a) to describe a colour without using colour names: 'light orange' or 'off-white' may suggest different colours to different people, whereas a Munsell notation defines one exact colour;

(b) to compare and select colours accurately:

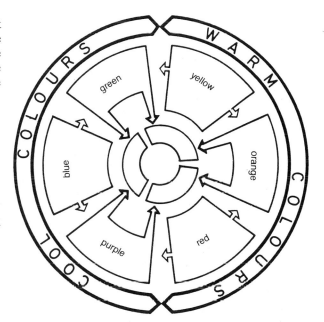

Fig. 8.3 *The colour circle*

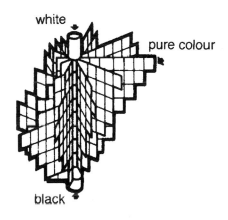

Fig. 8.4 *The Munsell-scale atlas*

the hundreds of colours defined by the Munsell scale allow the correct colour for a particular purpose to be chosen easily.

The full range of colours is illustrated in an atlas which can be used to select and specify colours.

The Munsell scale of colours has been used to prepare BS 4800:1994 and BS 5252: 1996 which give standard colours for decorative paints. A list of approximate Munsell references for each of the colours in the range is published by the British Standards Institution.

The system is based on the three properties of *hue, value* and *chroma*.

Hue (Fig. 8.5)
Hue denotes the basic colour, distinguishing yellow from red, blue from green, etc. Munsell uses ten principal hues:

yellow (Y)	purple-blue (PB)
yellow-red (YR)	blue (B)
red (R)	blue-green (BG)
red-purple (RP)	green (G)
purple (P)	green-yellow (GY)

Each principal hue is subdivided into ten sections, providing a full circle of 100 hues. The true hue is always prefixed by a 5, e.g. 5Y represents pure yellow. 7.5Y is two and a half divisions away from pure yellow in the blue direction, therefore slightly greenish; 2.5Y is the other side of pure yellow and is slightly reddish.

Value (Fig. 8.6)
Value measures the lightness and darkness of a colour. The scale is based on a vertical pole divided into eleven divisions. White is 10 and black is 0 and the remaining nine divisions represent a scale of greys from light (9) to dark grey (1). The value scale is shown by a number between 1 and 9 written to the right of the hue, e.g. pure yellow will be written as 5Y8 because the tone of pure yellow is equivalent to 8 on the value scale.

Fig. 8.5 *Hue*

Fig. 8.6 *Value*

Chroma (Fig. 8.7)

Chroma means the greyness of a colour. It is measured on a horizontal scale from neutral grey (0) to the pure hue which may be 10, 12 or 14 depending on the strength of the colour, e.g. pure PB is chroma 10 whereas pure Y is chroma 14. The chroma range number is written after the value scale and separated from it by an oblique stroke, e.g. pure yellow is written as 5Y8/14, whereas this colour darkened with black and reduced in strength by the addition of grey may be described as 5Y6/6.

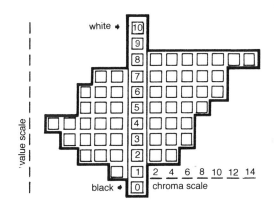

Fig. 8.7   *Chroma*

8.3 COLOUR STANDARDS

8.3.1 BS 1710:1991 SPECIFICATION FOR IDENTIFICATION OF PIPELINES AND SERVICES

This standard specifies a colour coding system for use on pipes and conduits in land installations and on board ships. Each colour identifies a particular liquid, gas or electrical service. The system ensures that even in a complex pipework system, in the case of fire, other emergency or maintenance work, the contents of every pipe can be easily identified and traced to source.

Where it is sufficient to identify the general nature of the contents, basic identification colours (see Table 8.1) are used. Where more precise information is required, code indications are added according to regulations.

Basic identification colours can be applied either by painting the whole length of the pipe in the specified colour, or by placing 150 mm wide bands of paint or coloured adhesive tape at all junctions, on both sides of valves, service appliances and wall penetrations or wherever identification is necessary.

8.3.2 BS 4800:1994 SCHEDULE OF PAINT COLOURS FOR BUILDING PURPOSES

The standard range contains 100 colours chosen from a basic range of 237 colours established in BS 5252:1996 *Framework for colour co-ordination for building purposes*. The range is divided into a Basic selection of 32 colours, including black and white, and two Supplementary selections relating to oil based finishes, category G, and emulsions, category M. The Basic selection are those colours which should be available from stock. Each colour is identified by a code consisting of three parts.

TABLE 8.1 **Basic identification colours for pipelines**

Pipe content	Colour name	Approx BS 4800 colour reference
Water	Green	12 D 45
Steam	Silver-grey	10 A 03
Mineral vegetable and animal oils: combustible liquids	Brown	06 C 39
Gases and liquid gases (except air)	Yellow ochre	08 C 35
Acids and alkalis	Violet	22 C 37
Air	Light blue	20 E 51
Other fluids including drainage	Black	00 E 53
Electrical services	Orange	06 E 51
Fresh water	Auxiliary blue	18 E 53

Hue

The first part signifies hue or colour and consists of an even number of two numerals, e.g. 04. Twelve main hues are used,

numbered

02	red purples
04	reds
06	yellow reds
08	yellow reds
10	yellows
12	yellow greens
14	greens
16	green blues
18	blues
20	purple blues
22	violets
24	purples

plus greys which are all 00.

Greyness

The second part signifies greyness, i.e. the difference in the apparent amount of greyness in one colour compared with another. Five grades are used, each defined by a single letter. There are four steps of diminishing greyness from A maximum to D minimum. Beyond this, colours are pure or free from greyness, graded as E: e.g. pure yellows are prefixed 10 E.

Weight

When coded only on the above two grades it was found that the yellow hues (yellow-red and yellow-green) at minimum greyness looked heavy in comparison with the other colours. This was overcome by raising the value of the yellowish colours. The result was more uniform in weight. The weight is given in pairs of numbers from 01 to 56.

Groups of colours within each of the five greyness ranges are graduated from high value to low value. Each of these graduation is numbered:

A greyness – 01 to 13
B greyness – 15 to 29
C greyness – 31 to 40
D greyness – 43 to 45
E greyness – 49 to 56

E.g. the lighter yellow is 10 E 49 while the deeper pure yellow is 10 E 53.

8.3.3 BS 5378 : 1995 (Part 1) SAFETY SIGNS AND COLOURS

This standard specifies an international system that gives safety information with the minimum use of words. It defines safety colours and safety signs (see Tables 8.2 and 8.3) for giving messages for use in

(i) Prevention of accidents.
(ii) Identification of health hazards.
(iii) Meeting emergencies.

Note All safety signs must conform to these standards.

Definitions

Safety colour A colour which has a special safety meaning or purpose.
Safety symbol A graphic symbol used in a safety sign.
Safety sign A sign which states a safety message by combining a safety colour with a safety symbol or both on a geometrical shape.
Supplementary sign A sign with a text only.

TABLE 8.2 **Safety signs**

Safety sign	Layout of signs	Example of use
	Prohibition signs White background with red circular band and cross bar. Safety symbol in black, placed centrally. Must not obliterate the cross bar. Sign should be at least 35 per cent red.	 SMOKING PROHIBITED
	Warning signs Yellow background with triangular band in black. Safety symbol or text in black, placed centrally. A black and yellow identification sign can be used where there is a risk of hazards. Sign should be at least 50 per cent yellow.	 CAUTION RISK OF FIRE RISK OF HAZARD
	Mandatory signs Blue circle. Safety symbol or text white, placed centrally. Sign should be at least 50 per cent blue.	 RESPIRATORY PROTECTION MUST BE WORN
	Emergency signs Green square or rectangle. Safety symbol or text in white, placed centrally. Sign should be at least 50 per cent green.	 FIRST AID
	Supplementary signs White square or rectangle. Text in black. Alternatively the background can be the same safety colour as the safety sign it is supplementing with the text in the appropriate contrasting colour.	FIRE EXIT

TABLE 8.3   **Safety colours**

Colour	Meaning or objective	Example of use	Contrast colour
Red 04 E 53	Stop Prohibition	Stop signs Emergency signs Prohibition signs such as NO SMOKING Fire-fighting equipment and its location	White
Yellow 08 E 51	Caution Risk of danger	Indication of hazards (fire, explosion, radiation, chemicals) Warning for thresholds, low passages, obstacles, projections	Black
Green 14 E 53	Emergency signs Safe conditions	Escape routes Emergency exits Emergency showers First aid Rescue stations	White
Blue 18 E 53	Mandatory action Information	Obligation to wear safety equipment Location of telephone	White

8.4 TERMS RELATED TO COLOUR SCHEMES

Achromatic colours A range of colours without hue: a series of greys ranging from black to white.

Advancing colours Warm colours taken from the red, yellow or orange part of the colour circle. They give the appearance of coming out or advancing towards the eye.

Analogous colours (Fig. 8.8) Harmonious or pleasing colours closely related to each other on the colour circle, e.g. yellow, yellow-green and green.

Fig. 8.8 *Analogous colours*

Complementary colours (Fig. 8.9) Any two colours which lie opposite each other on the colour circle and when mixed will produce grey or a neutral colour.

Fig. 8.9 *Complementary colours*

Contrasting colours Colours different in hue, value or chroma when seen together, e.g. the contrast of red against green, light colours against dark colours, strong colours against greyed colours.

Contrasting harmony Colours taken from various parts of the circle which appear pleasant when used together.

Cool colours Colours from the green and blue part of the colour circle, which given the appearance of coolness.

Discordant colours Colours used out of their natural tonal order (see Natural order of colour) e.g. pale blue with deep orange.

Harmonious colours Colours which appear pleasant when used together in a scheme. Common combinations are:
(a) Monochrome
(b) Analogous
(c) Contrasting
(d) Split complementary.

Juxtaposition of colour The effect one colour has on another when viewed side by side. The complementary colour of each will intensify or distort the effect of its neighbour: e.g. when red and yellow are seen together, the red will take on a purplish tinge while the yellow will appear greenish.

Fig. 8.10 *Monochrome*

Monochrome or dominant hue (Fig. 8.10) A colour scheme based on tints and shades of one colour.

Natural order of colour Yellow is the lightest colour in the colour circle and purple-blue the darkest. Used in this order they are usually pleasing to the eye. Reverse this order and colours become discordant.

Pastel colours Colours which have black and white added. Usually refers to pale colours or tints with a predominance of white added.

Retiring colours Cool colours taken from the blue-green part of the colour circle. They give the appearance of receding or retiring away from the eye (opposite to advancing).

Fig. 8.11 *Split complementary colours*

Shades Colour with black added.

Split complementary (Fig. 8.11) A colour scheme based on one hue contrasted with two hues either side of its complementary colour.

Tints Colours with white added.

Tones The lightness or darkness of a colour, i.e. the value rating on the Munsell scale.

Warm colours Advancing colours from the red-yellow-orange part of the colour circle, which given the appearance of warmth.

PART NINE

Glazing

9.1 Types of glass and glazing compound 279

9.1.1 Clear sheet glass 279
9.1.2 Patterned glass 279
9.1.3 Safety glazing materials 280
9.1.4 Glazing compounds 280

9.2 Handling and storage of glass 281

9.2.1 Handling glass 281
9.2.2 Storage 281

9.3 Measuring and cutting glass 282

9.3.1 Measuring for glass 282
9.3.2 Planning for cutting glass 282
9.3.3 Relevant glazing sizes 282
9.3.4 Cutting glass 282

9.4 Methods of glazing 283

9.4.1 Timber frames 283
9.4.2 Metal frames 283
9.4.3 Setting or location blocks 283

9.5 Double glazing 284

9.1 TYPES OF GLASS AND GLAZING COMPOUND

Glass is available in many types, grades, qualities, colours and sizes. Two main groups are used in domestic and industrial buildings and it is from these groups that the decorator may be called on to replace small broken panes of glass:

(a) Transparent glass that allows light to pass through it and that can also be seen through, i.e. plate glass or clear sheet glass.
(b) Translucent glass that allows light to pass through it but cannot be seen through, i.e. patterned glass.

9.1.1 CLEAR SHEET GLASS

Drawn sheet glass is available in the following qualities:

OQ (Ordinary glazing quality) Suitable for general glazing purposes.
SQ (Selected glazing quality) For work requiring glass above the ordinary glazing quality.
SSQ (Special selected quality) For high-grade work where superfine quality is required.
Horticultural 3 mm lower-grade glass for greenhouse work.

Float glass
The most common flat sheet glass. The liquid glass is floated on a bath of molten tin to produce clear glass of regular thickness, with few imperfections.

Available in two qualities only:
G.C. (General glazing quality) Suitable for shop fronts, showcases, shelves and table tops.
S.G. (Selected glazing quality) Suitable for work where specially high standards are required such as large areas of mirror.

9.1.2 PATTERNED GLASS

There are many types of patterned glass available in nominal thickness of 3 or 5 mm. They have two main uses:

(a) To provide a decorative effect in doors, screens and partitions.
(b) To provide privacy in rooms such as bathrooms and toilets and for partitions and screens. A clear image cannot be seen through them, but the degree of privacy depends on the type and depth of the pattern. They are classified in five groups graded from A to E, A being the clearest and E the most obscure.

TABLE 9.1 **Grades of clear sheet glass**

Nominal thickness mm	Approximate weight kg/m²	Maximum area Glazed m²	Use
4	9.5	0.2	Glazing small areas in domestic buildings
5	12.5	0.8	
6	15.0	1.8	Glazing large areas, mirrors, shelves etc.
Thicker glass available			
10		3.3	
12		5.0	

Note: Not less than 6 mm annealed glass should be used in doors and side panels.

9.1.3 SAFETY GLAZING MATERIALS

To ensure safety in the use of glass, all builders, architects, and specifiers need to comply with the safety section of BS 6262: 1982 *Code of practice for glazing in buildings*, which isolates risk areas in buildings and sets out minimum requirements to ensure safety.

The code recommends that where many people, especially children, are moving about, glazing situated lower than 800 mm from floor level should be of a safety glazing material of which there are four main types:

Wired Glass Clear or patterned glass enveloping a fine wire mesh. Sometime called fire resistant glass as it will act as a shield against the spread of flames for up to 30 minutes, depending on the size and type of frame. If during a fire the glass breaks, the wire holds the glass in position.

Laminated glass Two or more sheets of ordinary glass between which are inserted interlayers of a tough, resilient plastic material or resin. When broken, the fragments remain in place, adhering to the plastic.

Toughened glass Tempered glass. Four to five times stronger than ordinary glass, requiring considerable impact to break. When broken, it disintegrates into small, relatively harmless pieces. *Note* toughened glass cannot be cut and must be ordered in exact sizes required.

Glazing plastics Usually acrylic, polycarbonate, or rigid PVC. They all resist considerable impact and are not dangerous when broken.

9.1.4 GLAZING COMPOUNDS

Four main types of glazing compound are used.

Linseed oil putty

Made from linseed oil, whiting and a special filler. It hardens by oxidation. Used for bedding and fronting in primed timber frames on domestic and industrial work including roof work. Needs to be protected by painting within two weeks, after the surface has hardened sufficiently to take the paint. If required, it should be softened with linseed oil only.

Metal casement putty

Made from linseed and other drying oils, and a special grade of whiting. It hardens by oxidation. Used for bedding and fronting glass in metal and concrete frames on domestic and industrial work including roof work. Needs to be painted within the period specified by the manufacturer – usually 7–28 days. If required, it should be softened with special thinner supplied by the manufacturer.

Glazing compounds

Made from drying and non-drying oils, whiting and usually butyl rubber. They set but remain permanently flexible. Used on timber frames with face beads, concrete frames and metal bead work. Need to be painted within a period specified by the manufacturer. If required, should be thinned with special thinner supplied by the manufacturer.

Multi-purpose compound

Formulated on calcium carbonate fillers and selected oils. Intended for 'face glazing' of primed softwood/hardwood and steel window frames.

9.2 HANDLING AND STORAGE OF GLASS

9.2.1 HANDLING GLASS

When glass is being carried, the following points should be observed:

(i) It is safer to carry more than one pane at a time as this prevents the sheet whipping (Fig. 9.1 a).

(ii) Always make sure that the path ahead is clear and free from scaffolding, low doors and other hazards (Fig. 9.1 b).

(iii) Never stop suddenly or step backward: anyone following behind may be injured (Fig. 9.1 c).

(iv) Small panes of glass should be carried balanced centrally under the arm (Fig. 9.1 d).

(v) Larger panes should be carried with one hand underneath and the other in the front (Fig. 9.1 e).

(vi) Glass should be carried almost vertical and close to the body (Fig. 9.1 f).

9.2.2 STORAGE (Fig. 9.2)

Glass can easily be broken if not stored correctly, and the face of glass can be damaged if it

Fig. 9.2 *Storage of glass*

is stacked damp. The following rules should be observed:

(i) Never store lying flat: glass should be stood on one long edge, on timber or felt blocks, at an angle of approximately 80° or almost upright.

(ii) There should be no spaces between sheets; this will cause warping and the glass will eventually break.

(iii) Glass must be stored dry and protected from the wind – indoors if possible.

Fig. 9.1 *Handling glass*

9.3 MEASURING AND CUTTING GLASS

Figure 9.5 shows the names of various sizes referred to in glazing. *Tolerance* is the amount of variance allowed when measuring and cutting glass, usually $\pm$ 1 mm on the glazing size: e.g. a height size of 360 mm should be not less than 359 mm and not more than 361 mm.

9.3.1 MEASURING FOR GLASS

(i) When measuring for any type of glass it is standard practice to give the width size first. This ensures that when decorative glass is being used the pattern is always glazed the correct way, e.g. pattern vertical.

(ii) If the frame is unsymmetrical or oddly-shaped a small sketch should be made and the front marked *face* (Fig. 9.3). This ensures that the glass is cut correctly to fit it.

Note All glazing sizes should be expressed in millimetres (mm).

Fig. 9.3   *Measurement of glass*

9.3.2 PLANNING FOR CUTTING GLASS (Fig. 9.4)

Before attempting any cutting, an economic plan must be decided upon to eliminate as much waste as possible. The required size should be considered in conjunction with the size of sheet available.

Well planned cutting should produce strips of waste glass not exceeding 50 mm to 100 mm wide.

Fig. 9.4   *Planning for cutting glass*

9.3.3 RELEVANT GLAZING SIZES (Fig. 9.5)

T/S (tight size)
The actual full measurement between the glazing bars.

S/S (sight size) or (daylight size)
The measurement of space between the glazing bars which allows light through.

G/S (glazing size)
The actual measurement of the glass allowing for the edge clearance of 2 mm all round.

Fig. 9.5   *Glazing sizes*

9.3.4 CUTTING GLASS

Glass should always be cut on a flat, rigid, padded surface. If a glazier's felt covered cutting bench is not available, a flat surface padded with newspaper will suffice.

9.4 METHODS OF GLAZING

9.4.1 TIMBER FRAMES

The two main methods of glazing timber frames are (a) with putty or glazing compound and (b) with beads.

Putty glazing (Fig. 9.6)
The timber frame, which must be clean and primed, is bedded with a 'back putty'. The glass, cut with a 2 mm edge clearance, is fitted and small sprigs (small headless nails) are tapped in at intervals of approximately 460 mm around the frame. These hold the glass in until the putty sets. The front putty is then run and faced at an angle to provide a weather-proof fillet.

Fig. 9.6 *Putty glazing*

Glazing with beads (Fig. 9.7)
The timber frame must be clean and primed. The back putty is run, the glass fitted and the beads are bedded and fixed with either screws or panel pins.
Note Hardwood frames should be either primed or sealed with exterior quality varnish, and non-setting compounds used for glazing.

Fig. 9.7 *Glazing with beads*

Fig. 9.8 *Glazing metal frames*

9.4.2 METAL FRAMES (Fig. 9.8)

This process is similar to glazing timber frames, but with the following differences:
  (i) Metal casement putty is used.
 (ii) A metal primer is used.
(iii) The glass must always be cut with an edge clearance of 2 mm all round to allow for expansion and contraction of the metal and to allow for the metal clips. Otherwise the glass may be pinched and may crack.
(iv) In place of sprigs, spring clips are used to hold the glass in position. One end of the clip is fitted into small holes in the metal frame, the other end is sprung against the glass.

9.4.3 SETTING OR LOCATION BLOCKS (Fig. 9.9)

Panes of glass exceeding 0.2 m^2 should be set on blocks (often PVC) to ensure that the glass fits squarely into the frame (Fig. 9.9).

Fig. 9.9 *Setting blocks*

9.5 DOUBLE GLAZING

Double glazing consists of two panes of glass with a space ranging between 6 mm and 20 mm. This provides better sound and heat insulation.

 (i) sound insulation, to reduce the amount of noise entering or leaving a building.
(ii) heat insulation, to reduce the amount of heat energy escaping from a building.

Double glazing

Double glazing is where a patent double glazed unit is fitted as standard.

Secondary double glazing is where an additional layer is added to an existing single glazed component.

Types

Double window Two single glazed timber or metal frames, spaced apart in the same wall opening and independently mounted.

Coupled windows A single glazed frame fitted with an auxiliary frame coupled to it, allowing both to move together. Can be separated for cleaning purposes.

Dual glazing A secondary frame of metal, plastic, or wood fitted to an existing frame. Fitted on slides or hinges and can be separated for cleaning and maintenance. Can also be screwed down as a more permanent structure.

Factory made sealed units Aluminium or uPVC frames containing hermetically sealed double glazed units. Prevents the infiltration of dust or water vapour into the cavity between the glass.

INDEX

Note In addition to the Contents, each of the nine parts of this book is preceded with a detailed list.

If the required reference is not found in the Index, refer first to the Contents, and then to the appropriate lists preceding the book part.

Abrasives, 129–31
  for blast cleaning, 35
Absorbency, of building materials, 138
AC current, 58
Acetylene, 61
Acid pickling, 146
Acoustic board, painting of, 141
Air brush, 55
Air caps, 39 41
Air compressor
  air delivery of, 48
  types of, 44
Air consumption of spray guns, 39, 48
Air delivery
  of compressors, 44, 48–50
  definition of, 58
Air displacement, 58
Air pressure
  definition of, 57
  spray gun requirements of, 48–50
Air transformer, 46
Air volume, 58
Airless spray, 51
  advantages of, 53, 247
  fluid hose for, 47
  fluid tip, 52
  maintenance of, 54
Algae, 151
Alkalis, 144
Alternating current, 58
Aluminium leaf *see* Gilding
Aluminium oxide
  abrasive paper, 129–30
  grit, 35
Anode *see* Corrosion
Architect, 255, 257
Artificial light, 266
Asbestos, Regulations *see* Regulations
Asbestos cement, preparation of, 235
Asphyxia, 235, 239

Atomisation, 58

Back priming, 141
Badger hair, 14
Ballotini spheres, 94
Barrier cream, 230
Belt sander, 10
Bill of quantities, 252, 257
Blast cleaning, 34–7
  air consumption of, 57
  of metals, 146
Bleach, 151
Bleaching
  of paint, 153
  of wood, 149
Bleeder gun, 38
Blockboard, painting of, 142
Bottom outlet tank, 44
Brickwork, painting of, 144
Bristle, 13, 14
British Standards, 261
  BS EN Standards, 262
Broken colour texturing
  materials for, 105–6
  surface preparation for, 157
  tools for, 22
Bronze powders *see* Bronzing
Bronzing
  materials for, 113–4
  surface preparation for, 157
Brush painting, advantages of, 253
Building blocks, painting of, 144
Burning off, 135
  equipment for, 11, 12
  hazards of, 224, 229
Butane gas *see* LPG
Button polish, 105

Carbon monoxide gas, 63, 232
Casing wheel, 20

Cathode *see* Corrosion
Caulking, 132
    tools for, 3, 7
Caustic soda
    as a paint remover, 135
    protection against, 136, 233
Cement rendering, painting of, 144
Chipboard, painting of, 143
Chroma, 269
Circuit breaker *see* Residual current circuit
        breaker
Clear finishes, 101, 149
Coalescence, 68
Coal tar oils, 101
Cold cured epoxies, 95
Cold galvanising, 84
Compressed air, 57
Concrete
    painting of, 144
    papering of, 161
Construction Regulations *see* Regulations
Convertible coatings, 68
Corrosion, 145, *see also* Rust

DC current, 58
Dead hand *see* White finger
Decibel levels, 236
Defects
    of gilding, 115
    of paint, 107, 153
    of surfaces, 151
    of wallpapers, 178
Degreasing
    of metals, 146
    of sound surfaces, 136
Dehydration, of rust, 33, 146
Dermatitis *see* Skin disease
Descaling tools, 10–11
    air consumption of, 49
    comparisons, 146
Detergent
    as a washing material, 137
    hazards in use of, 230
Diaphragm compressor, 45
Diesel engine, 61–2
Differential expansion of metals, 33
Diluent *see* Thinners
Dim book, 255, 257
Direct current, 58
Driers, 67, 77
    excessive use of, 154
Dry lining, 141
    papering of, 161
Dry rot *see* Rot

Drying oils, 72
Drying processes of paint, 68
Dust
    hazards of, 235
    protection against, 233
Dutch metal *see* Gilding

Ears *see* Noise
Electricity, 58
    single and three phase, 59
    regulations, 220
Electric shock
    cause of, 59
    first aid for, 240
Electrostatic spray, 51
Emery abrasive, 129
    use of, 147
Estimate, 248, 250
Evaporation, 68
Expanded polystyrene
    as a wall covering, 164, 166
    properties of, 143
Extenders, 71
Eyes
    hazards to, 235
    protection of, 234
    regulations, relating to, 221

Ferrous metals
    definition of, 145
    priming of, 146
Fibre brush filling, 14
Fibreboard, 142
Film formers, 67–76
Finishing sander *see* Orbital sander
Fire, cause of, 223
Fire proofed boards, 145
Flame cleaner, 33
    use of, 146
Flame spread
    classification of, 140
    reduction of, 95
Flash point, 77, 223
Flock, 56
Flocking gun, 56
Floor paint, 95
Fluid hose, 46
Fluid pump, 51
Fluid tip, for airless spray, 52
Four stroke engine, 62
Fragile roofs, working on, 191
French polish, 105
Fumes
    cause of, 232

protection against, 233–5
Fungus *see* Moulds
Furs, brush filling, 14

Galvanic series, 147
Galvanising, 147
Garnet paper, 129
Gases
   hazards of, 232
   properties of, 61
   storage of, 227
Gas torch, 11
Gilding, 112
   defects of, 115
   materials for, 112–3
   tools for, 29–30
Glass
   painting of, 143
   types of, 279
   see also Part Nine **Glazing**
Glass reinforced cement board, 144
Glaze, oil, 105
Glazed tiles, painting of, 143
Glazing
   methods of, 283
   tools, 4, 7
   Gold leaf *see* Gilding
Gold size, 114
Graining
   materials for, 105–6
   surface preparation for, 156
   tools for, 22
Gravity feed container, 42
Gypsum plaster
   classes of, 144
   painting of, 144
   papering of, 161

Hand cleansers *see* Skin cleansers
Hardboard
   painting of, 142
   papering of, 161
Hardwood
   painting of, 142
   transparent coating of, 149
Head protection, 219
Health and Safety Executive, 217
Health and Safety at Work Act, 217
HEPA filters, 235
Highly flammable liquids,
   regulations covering storage, 223, 226
Highways Act, 213
Hop-ups, 193
Hopper feed gun, 56

Horsehair, brush filling, 14
Hot air paint stripper, 12
Hot spray, 50
Hue
   BS 4800, 270
   Munsell, 268

Industrial gloves
   types of, 230
   uses of, 131, 233
Inertia anchor, 211
Insulating board
   painting of, 140
   papering of, 161
Internal combustion engine, 61–3
International wiring code, 60
Intumescent paint, 95
Isocyanate, 227

Jobbing 255
Jointing tape
   use in caulking, 132
   use in papering, 162

Knotting, composition of, 104

Lambswool, 18
Lead paint coatings
   hazards of burning, 229
   regulations, 229
Lichen *see* algae
Lifting
   regulations, 221
   safe methods of, 237
Light
   artificial, 266
   natural, 266
Lime plaster
   painting of, 144
   papering of, 161
Lining
   tapes, 125
   tools and brushes, 27
Liquefied petroleum gas *see* LPG
Liquid paint remover *see* Spirit paint remover
Long oil varnish, 103
LPG
   properties of, 60–61
   storage of, 218
   use in paint burners, 11
LPG torch, 11

Marbling
   materials for, 105–6

surface preparation for, 156
  tools for, 22
Masking materials, 116–7
Medium density fibreboard (MDF) Fibreboard
Mediums, for gilding and bronzing, 114
Metals, effective paint system, 146
Microporous coatings, (MVP), 90, 101
Millscale, 145
  equipment for removal of, 11, 33–4
Mohair, for rollers, 18
Moisture content, measurement of, 141
Moisture meter, 141
Mordants, for gilding, 114
Moss *see* Algae
Mould growth, 151, 162, 181

Needle gun, *see* Descaling tools
Noise
  causes of, 235
  protection against, 234
  regulations, 221
Non-bleeder gun, 39
Non-convertible film, 68
Non-ferrous metals, painting of, 147
Non-intumescent paint, 95
Non-reversible film, 68

Oil modified resins
  natural, 75
  synthetic, 73
Oil stain
  for wood, 102
One stroke engine, 62
Open coated abrasives, 129
Orbital sander, 9
Outrigger, 205
Overalls, 225
Overheads, 257
Ox hair, 14
Oxalic acid, 149
Oxidation
  as a paint drying process, 68
  of drying oils, 75
Oxygen
  as a bottled gas, 61
  as an element in fire, 225
  see also Oxidation

Paint
  composition of, 67
  drying of, 68
  film thickness of, 146
  properties of, 77
  storage of, 226

types of finish, 69
Paint mitten, 19
Paint pads, 19
Painting, reasons for, 77
Paperhanging *see* Part Four **Surface Coverings**
Patent knotting, 104
Petrol
  engine, 61–3
  storage of, 227
Phosphor bronze tools, 9, 224
Pigments, 69–71
  effects of acids and alkalis upon, 153
Piston compressor, 44
Plaster *see* Gypsum plaster
Plasterboard
  painting of, 141
  papering of, 161–2
Plastics, painting of, 143
  dust hazard, 235
Plywood
  painting of, 142
  papering of, 161
Pneumatic tools
  air consumption of, 49–50, 57
  types of, 9–11
Polymerisation, 68
Polystyrene *see* Expanded polystyrene
Polyvinyl butyrl resin, 81
Porosity, of building materials, 138
Portable compressors, 45
Pot life
  of epoxy paint, 95
  of etch primer, 81
  of polyurethane paint, 97
Pressure drop, 47
Pressure feed containers, 43–4
  air consumption of, 49
Primary colours, 267
Profit, 257
Propane gas *see* LPG
Protection of eyes Regulations, provisions, 233
Protective clothing, 229
Pumice, 131

Quantity surveyor, 255
Quill brushes, 26
Quotation, estimate, 257

Raynaud's phenomenon, 236
Recovery position, 239
Regulations
  Asbestos, 217
  Electricity at work, 220
  First aid, 220

Handling, 221
Hazardous substances, 219
Head protection, 219
Health and safety, 220
Lead at work, 229
Noise, 221
Protective equipment, 221
Work place, 222
  see also Part 5 **scaffolding**
Remote cup, 43
  air consumption of, 49
Residual current circuit breaker, 59
Resinous wood
  paint defects on, 153
  painting of, 148
Resins, 67, 75
Respirators, 233
  regulations, 235
Resuscitation, 240
Reversible film, 68
Road marking/line paint, 94, 104
Roller painting
  advantages of, 253
  airless operated, 17
Rot
  dry, 151
  wet, 152
Rotary compressor, 45
Rotary disc sander, 8
Rotary wire brush, 9
Rust
  cause of, 145
  dehydration of, 33
  removal of, 33, 146
Rust inhibitive, pigments and primers, 146

Sable brush filling, 14
Sabox hair, 14
Sacrificial coating, 147
Safety colours, 271
Sanitary conveniences, provision of, 219, 222, 230
Scaffold
  parts of, 196–8
  tying in of, 196, 201
Scales, drawing, 251, 252
Schlag metal *see* Gilding
Scumble, 105
Secondary colours, 267
Shelf life
  of etch primers, 81
  of multi-colour paints, 88
Shellac knotting, 104
Shop priming, of wood, 142

Short oil varnish, 103
Signwriting tools, 26–8
Silicon carbide abrasive, 130
Silicone water repellant, 104
Silicosis, 235
Site facilities
  first-aid, 238
  hygiene, 230
SI units, 258
Skin cleansers, 230
Skin disease, 230
Sloping roofs, working on, 191
Softwood
  painting of, 142
  transparent coating of, 149
Solvents *see* Thinners
Spattering cap, 41
Specification, preparation of, 252–6
Spectrum, 265
Spirit paint remover
  hazards of, 231
  use of, 135
Spirit stain, 102
Split head trestle, 192
Spontaneous combustion, 224
Spray gun
  air consumption of, 39
  airless, 51
  cleaning of, 54
  lubrication of, 55
  paint containers for, 42–4
  storage of, 227
Spray painting, advantages of, 253
Squirrel hair, 14
Stainers, 69
Static compressor, 45
Steam stripper, 32
Steel wool, 131
Stencil materials, 116
Stonework, painting of, 144
Suction feed container, 42
Superglypta, 169
Synthetic
  brush fillings, 14, 15
  resins, 75
  roller coverings, 18
Syphon feed container, 42

Tertiary colours, 267
Texture relief painting, tools for, 30, 31
Thinners, types of, 76
  fumes from, 232
  properties of, 75
Thixotropy, 68, 77

Timber preservation *see* Wood preservatives
    hazards of dust, 235
Tower scaffold, 199
Tungsten carbide
    abrasive, 130
    spray tips, 52
Two stroke engine, 62

Universal scraper, 4
    use on wood, 149
Universal stainer, 69

Vacuum assisted strainer, 53
Value, 268
Veiling cap, 41

Wallpaper
    hand prints, 167

machine prints, 167
performance symbols, 185
sizes of, 184
terms for, 182
Water stain, 102
Wet rot *see* Rot
White finger, 236
White gold *see* Gilding
White light, 265
White polish, 105
Wood preservatives, 101, 149
Wood stain
    types of, 102
    uses of, 149
Woodwool slabs, painting of, 144
Working platform, 194